Studies in Systems, Decisi

Volume 363

Series Editor

Janusz Kacprzyk, Systems Research Institute, Polish Academy of Sciences, Warsaw, Poland

The series "Studies in Systems, Decision and Control" (SSDC) covers both new developments and advances, as well as the state of the art, in the various areas of broadly perceived systems, decision making and control–quickly, up to date and with a high quality. The intent is to cover the theory, applications, and perspectives on the state of the art and future developments relevant to systems, decision making, control, complex processes and related areas, as embedded in the fields of engineering, computer science, physics, economics, social and life sciences, as well as the paradigms and methodologies behind them. The series contains monographs, textbooks, lecture notes and edited volumes in systems, decision making and control spanning the areas of Cyber-Physical Systems, Autonomous Systems, Sensor Networks, Control Systems, Energy Systems, Automotive Systems, Biological Systems, Vehicular Networking and Connected Vehicles, Aerospace Systems, Automation, Manufacturing, Smart Grids, Nonlinear Systems, Power Systems, Robotics, Social Systems, Economic Systems and other. Of particular value to both the contributors and the readership are the short publication timeframe and the world-wide distribution and exposure which enable both a wide and rapid dissemination of research output.

Indexed by SCOPUS, DBLP, WTI Frankfurt eG, zbMATH, SCImago.

All books published in the series are submitted for consideration in Web of Science.

More information about this series at http://www.springer.com/series/13304

Joyjit Mukherjee · Indra Narayan Kar ·
Sudipto Mukherjee

Adaptive Robust Control for Planar Snake Robots

Joyjit Mukherjee ⓘ
SDU Robotics, The Mærsk Mc-Kinney
Møller Institute
University of South Denmark
Odense, Denmark

Indra Narayan Kar
Department of Electrical Engineering
Indian Institute of Technology Delhi
New Delhi, Delhi, India

Sudipto Mukherjee
Department of Mechanical Engineering
Indian Institute of Technology Delhi
New Delhi, Delhi, India

ISSN 2198-4182 ISSN 2198-4190 (electronic)
Studies in Systems, Decision and Control
ISBN 978-3-030-71462-8 ISBN 978-3-030-71460-4 (eBook)
https://doi.org/10.1007/978-3-030-71460-4

© The Editor(s) (if applicable) and The Author(s), under exclusive license to Springer Nature Switzerland AG 2021
This work is subject to copyright. All rights are solely and exclusively licensed by the Publisher, whether the whole or part of the material is concerned, specifically the rights of translation, reprinting, reuse of illustrations, recitation, broadcasting, reproduction on microfilms or in any other physical way, and transmission or information storage and retrieval, electronic adaptation, computer software, or by similar or dissimilar methodology now known or hereafter developed.
The use of general descriptive names, registered names, trademarks, service marks, etc. in this publication does not imply, even in the absence of a specific statement, that such names are exempt from the relevant protective laws and regulations and therefore free for general use.
The publisher, the authors and the editors are safe to assume that the advice and information in this book are believed to be true and accurate at the date of publication. Neither the publisher nor the authors or the editors give a warranty, expressed or implied, with respect to the material contained herein or for any errors or omissions that may have been made. The publisher remains neutral with regard to jurisdictional claims in published maps and institutional affiliations.

This Springer imprint is published by the registered company Springer Nature Switzerland AG
The registered company address is: Gewerbestrasse 11, 6330 Cham, Switzerland

Dedicated to Thakuma, Baba, Ma and Didi

विद्या ददाति विनयं विनयधाति पत्रताम् ।
पात्रत्वाद्धनमाप्नोति धनाद्धर्मं ततः सुखम् ॥

- हितोपदेशः

Knowledge gives humility; from humility one attains character; from character, one acquires wealth; from wealth follows good deeds and happiness.

-Hitopadesha

Preface

Snake robots are articulated systems designed to imitate the motion characteristics of biological snakes. These systems find a range of prospective applications due to their salient features like multi-modal locomotion and flexible motion capabilities. However, these features also pose challenges from the point of view of dynamical modeling and control. Hence, research in this field has found traction to make these systems worthy of field deployment.

The motion of a snake robot on a flat surface has been widely studied in literature. The state-of-art control approach for the planar snake robot employs a VHC-based body shape control along with singular perturbation-based head-angle control and feedback linearization-based velocity control. This approach requires the system to be known, which can be considered as a conservative assumption for unstructured and unknown environments. Moreover, the complicated dynamical system structure makes the control design problem challenging which has been conventionally solved through a multi-layer approach. Furthermore, to incorporate the versatility of snake motion in the robot, it is necessary to investigate the other modes of locomotion as well. Such studies would enhance the applicability of these systems in real-life scenarios.

In view of the discussion presented in Chap. 1 concerning the control problem of a planar snake robot and the state-of-art control approaches, this book presents four major research outcomes in Chaps. 2–5 enumerated as

- A Sliding-Mode Control (SMC) approach has been adopted to achieve efficient tracking for a planar snake robot considering bounded uncertainties in the friction coefficients with known upper bound. To circumvent the requirement of knowledge of the upper bound, an adaptive sliding-mode control (ASMC) law has been devised, which also alleviates the overestimation of the switching gain.
- The SMC and ASMC methodologies require the uncertainties to be bounded and are also dependent upon the friction model. A Time-Delayed Control (TDC) approach utilizing an artificial delay-based estimation, has been adopted to obtain satisfactory tracking efficiency of planar snake robot while obviating the aforementioned assumptions.

- The bounded estimation error in TDC has a detrimental effect on the tracking performance of a planar snake robot, which can only be reduced by choosing high feedback gain. For this reason, an Adaptive Robust Time-Delayed Control (ARTDC) law has been devised to achieve improved tracking performance in the presence of bounded estimation error. The novel dual-rate adaptation law is aimed at solving the overestimation problem of the switching gain.
- Differential flatness has been employed to establish mapping between the output and the input variables of the robot utilizing the serpenoid gait function. The flat system for a planar snake robot has been further utilized for feedback controller design. Subsequently, uncertainties have been considered in the flat system and an adaptive robust control law has been discussed for the same.
- The dynamic equations of motion of a snake robot inside a channel have been obtained through Newton-Euler formulation. The normal and traction forces generated through the interaction of robot link and channel wall have been modeled. A state-of-art control law has been employed to study the tracking performance of the robot inside a uniform channel. A flatness-based adaptive robust control law has also been employed to address uncertainties for the in-channel motion of a snake robot.

Overall, this book discusses the adaptive-robust control design of planar snake robots in an attempt to enhance their deployability in unknown environments.

Odense, Denmark
January 2021

Joyjit Mukherjee
Indra Narayan Kar
Sudipto Mukherjee

Contents

1	**Introduction**		1
	1.1 Why Snake Robots?		1
	1.2 Mechanics of Planar Snake Robots		3
		1.2.1 System Kinematics	4
		1.2.2 Friction Force Model	5
		1.2.3 System Dynamics	6
	1.3 Control Problem of Snake Robots		7
	1.4 Virtual Holonomic Constraint		8
		1.4.1 Virtual Holonomic Constraint for Euler–Lagrange System	8
		1.4.2 Virtual Holonomic Constraint for Double Pendulum	11
	1.5 Body-Shape Control		13
		1.5.1 Serpenoid Gait Function	13
		1.5.2 Virtual Holonomic Constraint-Based Control for Planar Snake Robots	15
	1.6 Output-Based Control		16
		1.6.1 Constraint System	16
		1.6.2 Head-Angle Control	18
		1.6.3 Velocity Control	19
	1.7 Toward a Practical Control Framework		19
		1.7.1 Robustness for Planar Snake Robots	20
		1.7.2 Multi-layered Control Methodology	20
		1.7.3 Modeling Other Modes of Propagation	21
	1.8 The Theme		21
	1.9 Organization of the Book		22
	References		23

2 Adaptive Sliding-Mode Control for Velocity and Head-Angle Tracking ... 27
- 2.1 Problem Formulation ... 28
- 2.2 Brief Outline of Sliding-Mode and Adaptive Sliding-Mode Control ... 28
  - 2.2.1 Sliding-Mode Control ... 29
  - 2.2.2 Adaptive Sliding-Mode Control ... 30
- 2.3 Sliding-Mode Control for Planar Snake Robot ... 31
  - 2.3.1 Sliding-Mode Control Law ... 32
  - 2.3.2 Stability Analysis ... 32
- 2.4 Adaptive Sliding-Mode Control for Planar Snake Robots ... 36
  - 2.4.1 Adaptive Sliding-Mode Control Law ... 36
  - 2.4.2 Stability Analysis ... 38
- 2.5 Simulation Results ... 40
  - 2.5.1 Simulation Scenario ... 40
  - 2.5.2 Results for Sliding-Mode Control ... 41
  - 2.5.3 Results for Adaptive Sliding-Mode Control ... 46
  - 2.5.4 Discussion ... 51
- 2.6 Summary ... 52
- References ... 53

3 Time-Delayed Control for Planar Snake Robots ... 55
- 3.1 Control Description ... 56
  - 3.1.1 Outer Layer Time-Delayed Control ... 57
  - 3.1.2 Inner Layer Time-Delayed Control ... 59
- 3.2 Stability Analysis ... 63
- 3.3 Simulation Environment and Results ... 65
- 3.4 Summary ... 70
- References ... 70

4 Adaptive Robust Time-Delayed Control for Planar Snake Robots ... 73
- 4.1 Control Description ... 74
  - 4.1.1 Outer Layer Adaptive Robust Time-Delayed Control ... 74
  - 4.1.2 Inner Layer Adaptive Robust Time-Delayed Control ... 78
- 4.2 Stability Analysis ... 80
  - 4.2.1 Stability Analysis of Dual-Adaptive Robust Time-Delayed Control ... 80
  - 4.2.2 Selection of Parameters ... 83
- 4.3 Simulation Results ... 84
  - 4.3.1 Discussion ... 85
- 4.4 Summary ... 91
- References ... 91

5 Differential Flatness and Its Application to Snake Robots 93
- 5.1 Brief on Differential Flatness 94
- 5.2 Flatness for Wheeled Mobile Robots 95
  - 5.2.1 Robot Kinematics 95
  - 5.2.2 Error Model of Robot Kinematics 96
  - 5.2.3 Error Model as Flat System 97
  - 5.2.4 Flatness-Based Control Law 98
  - 5.2.5 Stability Analysis 99
  - 5.2.6 Simulation Results 100
- 5.3 Flatness of Snake Robot Utilizing Serpenoid Gait 104
  - 5.3.1 Establishing Flatness 104
  - 5.3.2 Flat System 107
- 5.4 Feedback Control Law 108
  - 5.4.1 Stability Analysis 109
  - 5.4.2 Simulation Results 110
  - 5.4.3 Discussion 110
- 5.5 Adaptive Robust Control Design for Flat Systems 113
  - 5.5.1 Robust Control Law for Flat Systems with Uncertainties 115
  - 5.5.2 Adaptation Law 117
  - 5.5.3 Stability Analysis 118
  - 5.5.4 Simulation Results 119
  - 5.5.5 Discussion 120
- 5.6 Summary 124
- References 125

6 Modeling of In-Pipe Snake Robot Motion 127
- 6.1 Dynamic Modeling 128
  - 6.1.1 Contact Force Model 129
  - 6.1.2 Friction Force Model 133
  - 6.1.3 Moment Due to Contact and Friction Forces 134
  - 6.1.4 Dynamic Equations 135
  - 6.1.5 Serpenoid Gait Function 136
- 6.2 Conventional Control Methodology 136
  - 6.2.1 Body-Shape Control 137
  - 6.2.2 Head-Angle Control 137
  - 6.2.3 Velocity Control 138
  - 6.2.4 Simulation Results 138
  - 6.2.5 Discussion 139
- 6.3 Flatness-Based Adaptive Robust Control 142
  - 6.3.1 Flat System 146
  - 6.3.2 Adaptive Robust Control Law 147
  - 6.3.3 Adaptation Law 148

6.3.4 Simulation Results.............................. 149
6.3.5 Discussions................................... 150
6.4 Summary... 157
References... 158

7 Conclusions.. 159

Appendix A: Boundedness of the Time-Delayed Estimation Error for Outer Layer Time-Delayed Control.......... 161

Appendix B: Boundedness of the Time-Delayed Estimation Error for Inner Layer Time-Delayed Control.......... 163

Appendix C: Boundedness of Switching Gains for Adaptive Robust Time-Delayed Control............................. 165

Index... 167

Abbreviations and Notation

Abbreviations

ARC	Adaptive Robust Control
ARTDC	Adaptive Robust Time-Delayed Control
ASMC	Adaptive Sliding-Mode Control
BSC	Body-Shape Control
CM	Centre of Mass
SMC	Sliding-Mode Control
TDC	Time-Delayed Control
TDE	Time-Delayed Estimation
UUB	Uniformly Ultimately Bounded
VHC	Virtual Holonomic Constraints
WMR	Wheeled Mobile Robot

Symbols

$\mathbb{R}$	Set of Real Numbers
$\mathbb{R}^+$	Set of Positive Real Numbers
$\mathbb{R}^j$	Real space of dimension j
$\mathbb{R}^{j \times j}$	Real matrix of dimension $(j \times j)$
$\exists$	there exists
$\forall$	for all
$\mathbf{I}_{j \times j}$	Identity matrix of dimension $(j \times j)$
$\mathbf{0}_j$	Zero vector of dimension j
$\mathbf{0}_{j \times j}$	Zero matrix of dimension $(j \times j)$
rand(j)	Vector of dimension j with random values within $[0\ 1]$
$\Lambda > \mathbf{0}$	Positive definite matrix
$\Lambda < \mathbf{0}$	Negative definite matrix

$\lambda_{\min}(\Lambda)$	Minimum eigenvalue of matrix Λ
$\lambda_{\max}(\Lambda)$	Maximum eigenvalue of matrix Λ
$\|\Lambda\|$	Euclidean norm of matrix Λ
n	Number of links for snake robot
c_t	Tangential ground friction coefficient
c_n	Normal ground friction coefficient
m	Mass of each link
$2l$	Length of each link
J	Moment of inertia of each link
θ_i	Angle of ith link with respect to global X axis; $\boldsymbol{\theta} \in \mathbb{R}^n$
(p_x, p_y)	Position of the robot CM in global reference frame; $\mathbf{p} \in \mathbb{R}^2$
(x_i, y_i)	Position of the CM of ith link in global reference frame; $\mathbf{x}, \mathbf{y} \in \mathbb{R}^n$
τ_i	Torque applied on ith link by $(i+1)$th link; $\tau \in \mathbb{R}^{n-1}$
a	Semi-minor axis of elliptical link for in channel motion
b	Semi-major axis of elliptical link for in channel motion
d	Width of the channel
$(x_{i,\min}, y_{x_{i,\min}})$	Point on ith link with minimum abscissa
$(x_{i,\max}, y_{x_{i,\max}})$	Point on ith link with maximum abscissa
E_{rub}	Modulus of elasticity of link surface material
v_{rub}	Poisson's ratio of link surface material
ϵ_{rub}	Maximum virtual penetration of Link surface
$N_{i,l}$	Contact force applied by the left wall on ith link; $\mathbf{N}_l \in \mathbb{R}^n$
$N_{i,r}$	Contact force applied by the right wall on ith link; $\mathbf{N}_r \in \mathbb{R}^n$
$C_{f,c,i}$	Total Contact force on ith link along X direction; $\mathbf{C}_{f,c} \in \mathbb{R}^n$
$v_{i,l}$	Contact point velocity of ith link and left wall
$v_{i,r}$	Contact point velocity of ith link and right wall
$r_{i,l}$	Distance of ith link CM and its left wall contact point
$r_{i,r}$	Distance of ith link CM and its right wall contact point
$\mu_{i,l}$	Friction coefficient for ith link left wall contact
$\mu_{i,r}$	Friction coefficient for ith link right wall contact
$C_{i,l}$	Friction force applied by the left wall on ith link; $\mathbf{C}_l \in \mathbb{R}^n$
$C_{i,r}$	Friction force applied by the right wall on ith link; $\mathbf{C}_r \in \mathbb{R}^n$
$C_{f,y,i}$	Traction force on ith link along Y direction; $\mathbf{C}_{f,y} \in \mathbb{R}^n$
$C_{f,\tau,i}$	Moment by Contact and Traction force on ith link; $\mathbf{C}_{f,\tau} \in \mathbb{R}^n$
v_{rub}	Poisson's ratio of link surface material
ϵ_{rub}	Maximum virtual penetration of Link surface
$h_{x,i}$	Joint constraint force from ith link to $(i+1)$th link along global X direction; $\mathbf{h}_x \in \mathbb{R}^{n-1}$
$h_{y,i}$	Joint constraint force from ith link to $(i+1)$th link along global Y direction; $\mathbf{h}_y \in \mathbb{R}^{n-1}$

Chapter 1
Introduction

Abstract This book aims to address the robustness issues encountered in controlling the motion of a snake robot when moving on surfaces with varying ground conditions. The snake robot considered in this book is an articulated serial chain robot with multiple binary links connected through active joints. Classically, planar snake robots achieve translation by leveraging differential friction characteristics in different directions by creating undulations in its body. Hence, the quality of the contact and determination of the variables contributing to the contact and friction forces are crucial toward the efficient motion planning of a planar snake robot. Variation in the ground condition in particular will have a significant effect on the tracking performance of the robot which has to be dealt with through effective controller design. In this book, adaptive robust control techniques like Sliding-Mode Control (SMC), Adaptive SMC (ASMC), Time-Delayed Control (TDC) and Adaptive Robust TDC (ARTDC) have been employed to achieve robustness in tracking the performance of planar snake robot while trading off between performance, input effort and limits or determinism of parameters.

1.1 Why Snake Robots?

Researchers have drawn inspiration from nature, especially the animal world seeking ideas to design improved mechanisms. This approach of designing mechanisms inspired by living beings is called *Biomimetics*. Organisms worth exploring are *Snakes* which belong to the class *Reptilia* and primarily move by crawling or grazing their body over the surface below. This type of motion allows the snakes to move freely in almost any kind of terrain. The *versatile* motion characteristics of a snake have been classified as

- Multi-modal motion characteristics,
- Capability of tackling obstacles,
- Unparalleled ability to move through constrained spaces.

Snakes display flexibility to move through tight spaces, climb trees and even *jump* from one tree to another. Such motion characteristics reproduced with an effective mechanical system could allow the design of machines with wide accessibility, extensive reach and desirable flexibility. This initiated extensive research and study on intricacies of snake motion by first experimenting on biological snakes. Seminal work in this field involves the research in [1, 2] to study the motion characteristics of snakes. The two groups of frictional forces operating over the snake body: *ventral* friction which is between the ventral surface and ground and *lateral* friction which is between the sides of the body and the surface in contact has been discussed in [1]. Reference [2] studies the straight-line motion of the Argentinian Boa and explains the physiological phenomenon that propels the snake forward during rectilinear progression. Further research on biological aspects of snake locomotion and the versatile movements they exhibit has been reported in [3–6]. Anisotropic friction, kinematics and different modes of motion of various snakes like *Coluber constrictor* or eastern racers, *Crotalus cerastes* or horned rattlesnakes, *Nerodia fasciata* or banded water snake and *Pantherophis obsoletus* or western rat snake have been discussed in these works. More recently, [7] has explained the interaction of the muscles and body fluids to achieve the required motion for a swimming snake robot in water. The initial work regarding the motion of a snake from a mechanical perspective can be found in [8, 9] primarily aimed at synthesizing the planar motion of snake robot. Moreover, [8] has reported the basic kinematics of snake locomotion, the phenomenon of *sinus-lifting* for rapid movement often seen in some snakes and the adaptive nature of the gait pattern for straight movement, whereas [9] shows the speed analysis of some common snakes. A major breakthrough in this field was the proposal of a gait pattern called **Serpenoid Curve** [10–12] that describes a body shape function which can be adapted to control the motion. The gait function is a desired motion of the joints of the snake robot that results in a controlled motion of the whole robot. The snake robot developed in this work employed articulated bodies on passive wheels. This configuration doesn't allow the robot links to slip in the direction of the wheel axis resulting in motion tangential to the wheel. Subsequent works eliminated the wheels and utilized differential body-ground friction to obtain controlled motion. Analysis of a particular locomotion gait that can lead to *sidewinding* motion where the snake as a whole translates at an acute angle to the direction of wave propagation for hyper-redundant robots has been presented in [13], whereas [14] has discussed a kinematics based algorithm to obtain a gait pattern for a snake to move over an uneven surface. The undulatory motion of a snake robot has been studied from the geometric aspect in [15], whereas [16] has analyzed the oscillatory motion of such kinematic chains using Lie-Groups. The creeping motion of a snake-like robot which is the slowest and stealthiest mode of locomotion has been studied in [17]. The kinematic analysis of a snake robot has been detailed in [18], whereas the locomotion control of snake robot on a surface with uniform friction assumption has been studied in [19]. A chronological sequence of serpentine robots designed and realised is available in [20–24]. In this book, an essential summary of mechanical model of planar snake robots and the state of art of control strategies are available in Sects. 1.2, 1.5 and 1.6 sequentially.

1.2 Mechanics of Planar Snake Robots

Developing the mathematical model for the planar motion of an articulated snake robot gives an essential insight into the motion characteristics of the robot. The analytical model in particular facilitates in-depth analysis of the system and controller design. In this section, the dynamic equations of motion for a planar snake robot with n links joined through $(n-1)$ active joints have been presented. A schematic diagram of a snake robot in the $X-Y$ plane is shown in Fig. 1.1.

To present the analytical model of the planar snake robot compactly, utilization of additive and difference operands are convenient which can be written with the following matrix elements:

$$\mathbf{A}_{j,k} = \begin{cases} 1, & \text{for } j = k \\ 1, & \text{for } j+1 = k \\ 0, & \text{otherwise} \end{cases}, \quad \mathbf{D}_{j,k} = \begin{cases} 1, & \text{for } j = k \\ -1, & \text{for } j+1 = k \\ 0, & \text{otherwise} \end{cases}.$$

With matrix $\mathbf{A} \in \mathbb{R}^{(n-1)\times n}$ as the additive operator and $\mathbf{D} \in \mathbb{R}^{(n-1)\times n}$ the difference operator, they result in the addition and subtraction between consecutive elements of a vector, respectively, when multiplied with it. We also define

$$\mathbf{e} = \begin{bmatrix} 1 & 1 & \ldots & 1 \end{bmatrix}^T \in \mathbb{R}^n, \quad \mathbf{E} = \begin{bmatrix} \mathbf{e} & \mathbf{0}_{n\times 1} \\ \mathbf{0}_{n\times 1} & \mathbf{e} \end{bmatrix} \in \mathbb{R}^{2n\times 2},$$

where the vector $\mathbf{e}$ provides the scalar sum of all the element of a vector its operates on. Matrix $\mathbf{E}$ provides scalar sums of the first half and second of the operand vector as a two-dimensional vector. The vectors $\mathbf{b}_1$ and $\mathbf{b}_2$ operate to obtain the last element of a particular vector and can be expressed as

$$\mathbf{b}_1 = \begin{bmatrix} 0 & \ldots & 0 & 1 \end{bmatrix}^T \in \mathbb{R}^{n-1}, \quad \mathbf{b}_2 = \begin{bmatrix} 0 & 0 & \ldots & 1 \end{bmatrix}^T \in \mathbb{R}^n,$$

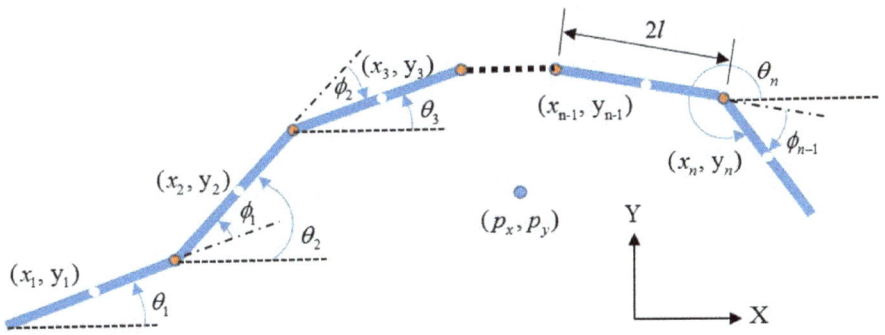

Fig. 1.1 Schematic diagram of a snake robot

$$\boldsymbol{\sin\theta} = \begin{bmatrix} \sin\theta_1 & \sin\theta_2 & \ldots & \sin\theta_n \end{bmatrix}^T \in \mathbb{R}^n \quad , \quad \mathbf{S}_\theta = \text{diag}(\boldsymbol{\sin\theta}) \in \mathbb{R}^{n\times n},$$
$$\boldsymbol{\cos\theta} = \begin{bmatrix} \cos\theta_1 & \cos\theta_2 & \ldots & \cos\theta_n \end{bmatrix}^T \in \mathbb{R}^n \quad , \quad \mathbf{C}_\theta = \text{diag}(\boldsymbol{\cos\theta}) \in \mathbb{R}^{n\times n}.$$

Vector $\boldsymbol{\sin\theta}$ and $\boldsymbol{\cos\theta}$ represents trigonometric operations on the angular positions of each link in the global frame. The vector containing the scalar squares of the each link's global angular velocity is given in

$$\dot{\boldsymbol{\theta}}^2 = [\dot{\theta}_1^2, \ldots, \dot{\theta}_n^2]^T \in \mathbb{R}^n.$$

Other matrices formed for a compact representation of kinematic and dynamic equation of motion are

$$\mathbf{N} = \mathbf{A}^T(\mathbf{DD}^T)^{-1}\mathbf{D} \in \mathbb{R}^{n\times n} \quad , \quad \mathbf{V} = \mathbf{A}^T(\mathbf{DD}^T)^{-1}\mathbf{A} \in \mathbb{R}^{n\times n},$$
$$\mathbf{H} = \begin{bmatrix} \mathbf{U}_{trig}(1)_n \\ \mathbf{0}_{1\times n} \end{bmatrix} \in \mathbb{R}^{n\times n} \quad , \quad \mathbf{SC}_\theta = \begin{bmatrix} \mathbf{N}^T\mathbf{S}_\theta \\ -\mathbf{N}^T\mathbf{C}_\theta \end{bmatrix}.$$

1.2.1 System Kinematics

A rigid body (say the head link of the snake robot) in a plane is specified by three variables: two position coordinates and an orientation. Successive links can then be located uniquely by specifying the articulation angles. Hence, it follows that if the output specification is only for the robot leading link, we specify 3 degrees of freedom. In this case, all joint actuated snake robots with more than a few links have redundant actuation and is amenable to an optimal control formulation [21, 25]. Define the angle that a link makes with the global positive X axis in an anti-clockwise sense as θ_i. Each joint angle in the sense of *D–H notations* can be expressed as the difference between the present and succeeding link angle. The frame of reference of each link is attached to its Center of Mass (CM) with its local X_i axis being along the length of the link, i.e. along the line connecting the axis of articulation and local Y_i axis is normal to the X_i direction in the plane. The rotation matrix aligning the ith link body frame to the global frame is expressed as

$$\mathbf{R}_{i\text{thframe}}^{\text{global}} = \begin{bmatrix} \cos\theta_i & -\sin\theta_i \\ \sin\theta_i & \cos\theta_i \end{bmatrix}.$$

The positional coordinate of the CM of the robot can be generally expressed as a weighted average of the CM coordinates of each link and is given by

$$\mathbf{p} = \begin{bmatrix} p_x \\ p_y \end{bmatrix} = \left[\left(\frac{\sum_{i=1}^n m_i x_i}{\sum_{i=1}^n m_i}\right) \left(\frac{\sum_{i=1}^n m_i y_i}{\sum_{i=1}^n m_i}\right) \right]^T.$$

1.2 Mechanics of Planar Snake Robots

Under the assumption of equal mass for all links ($m_i = m \ \forall \ i$), the robot CM coordinate in the global frame of reference can be expressed as

$$\mathbf{p} = \begin{bmatrix} \frac{1}{n}\sum_{i=1}^{n} x_i \\ \frac{1}{n}\sum_{i=1}^{n} y_i \end{bmatrix} = \frac{1}{n}\begin{bmatrix} \mathbf{e}^T \mathbf{x} \\ \mathbf{e}^T \mathbf{y} \end{bmatrix}. \tag{1.1}$$

From the geometry of the snake robot, the vector of positional coordinates of the links can be related to the vector of their angular positions in the global frame by the equality constraints written as

$$\begin{aligned} \mathbf{Dx} + l\mathbf{A}\cos\boldsymbol{\theta} &= 0, \\ \mathbf{Dy} + l\mathbf{A}\sin\boldsymbol{\theta} &= 0. \end{aligned} \tag{1.2}$$

For the case of the snake robot, define inverse kinematics as the process of solving for the positional coordinates of the link CMs given the link angles and global position of the robot CM. The inverse kinematic equations can be derived using (1.1) and (1.2) as [20, 21]

$$\begin{aligned} \mathbf{x} &= -l\mathbf{N}^T \cos\boldsymbol{\theta} + \mathbf{e}p_x, \\ \mathbf{y} &= -l\mathbf{N}^T \sin\boldsymbol{\theta} + \mathbf{e}p_y. \end{aligned} \tag{1.3}$$

The velocity of each each link CM can be obtained by differentiating (1.3) w.r.t. time and can be expressed as

$$\begin{aligned} \dot{\mathbf{x}} &= l\mathbf{N}^T \mathbf{S}_\theta \dot{\boldsymbol{\theta}} + \mathbf{e}\dot{p}_x, \\ \dot{\mathbf{y}} &= -l\mathbf{N}^T \mathbf{C}_\theta \dot{\boldsymbol{\theta}} + \mathbf{e}\dot{p}_y. \end{aligned} \tag{1.4}$$

The velocity equations can alternatively be represented in a vector form as

$$\Rightarrow \begin{bmatrix} \dot{\mathbf{x}} \\ \dot{\mathbf{y}} \end{bmatrix} = l\mathbf{SC}_\theta \dot{\boldsymbol{\theta}} + \mathbf{E}\dot{\mathbf{p}}. \tag{1.5}$$

1.2.2 Friction Force Model

The friction forces acting on the links in a plane are the only terms forming the propagating force of the robot. The total ground friction acting on the robot CM can be computed as the summation of the friction forces generated due to the undulation of each link. The friction forces acting on each link, when modeled using a viscous friction model, relating the force generated to the link velocities (1.4) by

$$\mathbf{f}_R(\boldsymbol{\theta}, \dot{\boldsymbol{\theta}}, \dot{\mathbf{p}}) = \begin{bmatrix} \mathbf{f}_{R,x} \\ \mathbf{f}_{R,y} \end{bmatrix} = \mathbf{Q}_\theta \begin{bmatrix} \dot{\mathbf{x}} \\ \dot{\mathbf{y}} \end{bmatrix} = l\mathbf{Q}_\theta \mathbf{SC}_\theta \dot{\boldsymbol{\theta}} + \mathbf{Q}_\theta \mathbf{E}\dot{\mathbf{p}}. \tag{1.6}$$

The velocities of each link CM in their respective body reference frames are mapped to the global frame of reference attached to the robot CM by the matrix

$$\mathbf{Q}_\theta = - \begin{bmatrix} c_t(\mathbf{C}_\theta)^2 + c_n(\mathbf{S}_\theta)^2 & (c_t - c_n)\mathbf{S}_\theta \mathbf{C}_\theta \\ (c_t - c_n)\mathbf{S}_\theta \mathbf{C}_\theta & c_t(\mathbf{S}_\theta)^2 + c_n(\mathbf{C}_\theta)^2 \end{bmatrix}, \qquad (1.7)$$

where c_t is the viscous friction coefficient in tangential direction and c_n is the same in normal direction to the axis of each link. By inspecting the friction equation, we see that the robot generates positive propagating force only when $c_n > c_t$. In biological snakes, the mode of locomotion called *Lateral Undulation* can be interpreted using this equation, where each link is worked upon by a normal and tangential friction force. The surface making contact with the ground is designed in such a way that the friction force along the normal direction is greater in magnitude than that in the tangential direction. This anisotropic friction is the primary reason behind the propagation of the snake robot [25].

1.2.3 System Dynamics

The articulation mechanism of a snake robot on hard ground is obtained when an oscillatory movement of equal amplitude is imparted to the articulated links with progressive phase delay between articulations. As the links of the robot execute this oscillatory movement, they interact with the ground below and generate propulsive friction force (1.6). The dynamic equations of the snake robot are given as [21, 25]

$$\mathbf{M}\ddot{\boldsymbol{\theta}} = \mathbf{W}\dot{\boldsymbol{\theta}}^2 + l\mathbf{SC}_\theta^T \mathbf{f}_R(\boldsymbol{\theta}, \dot{\boldsymbol{\theta}}, \dot{\mathbf{p}}) + \mathbf{D}^T \boldsymbol{\tau}, \qquad (1.8a)$$

$$\ddot{\mathbf{p}} = \left(\frac{1}{nm}\right) \mathbf{E}^T \mathbf{f}_R(\boldsymbol{\theta}, \dot{\boldsymbol{\theta}}, \dot{\mathbf{p}}), \qquad (1.8b)$$

where

$$\mathbf{M} = J\mathbf{I}_{n \times n} + ml^2 \mathbf{S}_\theta \mathbf{V} \mathbf{S}_\theta + ml^2 \mathbf{C}_\theta \mathbf{V} \mathbf{C}_\theta,$$
$$\mathbf{W} = ml^2 \mathbf{C}_\theta \mathbf{V} \mathbf{S}_\theta - ml^2 \mathbf{S}_\theta \mathbf{V} \mathbf{C}_\theta.$$

Balancing of angular moments about the CM of each link result in (1.8a). Equation (1.8b) can be obtained by equating the forces at the robot CM along the global X and Y direction. This equation represents the motion of the whole snake body in the global reference frame due to the friction forces generated as a result of the undulations.

1.3 Control Problem of Snake Robots

The complicated structure of the equations of motion for a snake robot moving on a planar surface makes the control problem challenging. The primary aim of any control framework is to efficiently maneuver the robot from an initial to a final state through the desired trajectory. In case of the snake system, we see that the states of the system representing the characteristic position of the robot are not directly influenced by the actuators which act at the robot joints.

To achieve propagation in all the methods reported so far, the joint variables of the robot execute a sustained oscillatory motion. Essentially, the controller has to drive these states to a stable limit cycle consistent with a time-varying body shape. The amplitude of undulations of the robot normal to the direction of motion should be controllable to generate the required thrust force. The head-link of the robot acts as a leader for the rest of the links and sets the trajectory which the other links follow. It is hence useful to transform the positional coordinates of the robot CM from the global frame of reference to the frame fixed to the head link. The position of the robot along the direction of motion is required to be tracked to a predefined value or function. This makes the control problem for a snake robot unique and challenging with a combination of limit cycle tracking and regulation problem.

Stabilization and trajectory control using the kinematic model for a practical snake robot has been discussed in [26]. A motion control technique using speed estimation to mimic the control method employed by snakes has been discussed in [27]. A control approach to track the head-link of a snake robot has been discussed in [28], whereas [29, 30] employs a dynamic manipulability-based methodology for the controlled locomotion of a wheeled snake robot where the head-link acceleration has been mapped to the normalized side force through a manipulability ellipsoid. The kinematic model of a snake robot has been utilized to design a control method through a redundancy controllable system approach to avoid obstacles as well as singular configuration that may result in a loss of controllability [31]. An active impedance control of the robot joints has been executed in [32] to achieve winding locomotion of robotic snake on rough terrain. An input–output linearization-based approach to track the joint angles of a snake robot to the serponoid gait function has been detailed in [33]. An iterative feedback control approach using principal fiber bundle modeling has been discussed in [34]. The controllability of a planar snake robot with anisotropic friction has been discussed in [21, 22] followed by the proposal of a multi-layer control approach through Virtual Holonomic Constraint (VHC)-based body shape control, singular perturbation-based head-angle control and feedback-linearization-based speed control in [35]. To address uncertainty in hydrodynamic forces for an underwater snake robot, an L1 adaptive controller with piecewise constant adaptive law has been proposed by [36], whereas [37] investigated the robustness of a redundant snake robot toward a single actuator failure during head-link trajectory tracking while also employing Kalman filter for state estimation. An adaptive control law based on the backstepping technique has been discussed in [38] to achieve desired motion for snake robots on surface with varying friction.

Variation in the environment, e.g. surface and hence friction changing as the robot transverses, may also demand adaptation and modification in the body shape of the snake robot. A Model Predictive Control (MPC) based approach has been proposed in [39] to compute optimal parameters for the gait function of a snake robot to execute the efficient motion. A Central Pattern Generator (CPG)-based framework has been presented by [40] to generate dynamic gait for the snake robot to tackle environmental uncertainties, whereas a decentralized control strategy has been discussed in [41] for controlling sections of the robot under the supervision of a meta-agent policy.

The work presented in this book focuses on the planar motion of an articulated snake robot on a flat surface. Various control approaches have been discussed to address robustness for the snake robot toward variation in ground conditions. Time-varying friction coefficient has been considered to imitate varying ground conditions. However, the conventional control approach for such systems has been discussed in detail in the subsequent sections to exemplify the motivation and contributions of the proposed methodologies.

1.4 Virtual Holonomic Constraint

Constraints that do not exist in the physical system, when induced artificially, are known to be effective in achieving a desired motion of the system [42, 43]. Such artificially induced constraints are called *Virtual Holonomic Constraints* (VHC) and are usually employed for establishing holonomic constraints between coordinates exhibiting oscillatory motion.

1.4.1 Virtual Holonomic Constraint for Euler–Lagrange System

Considering a dynamical system of the form,

$$\mathbf{M}(\mathbf{q})\ddot{\mathbf{q}} + \mathbf{C}(\mathbf{q}, \dot{\mathbf{q}})\dot{\mathbf{q}} + \mathbf{G}(\mathbf{q}) = \mathbf{B}(\mathbf{q})\tau, \tag{1.9}$$

where $\mathbf{q} \in \mathbb{R}^{\bar{n}}$ is the generalized coordinate vector and $\tau \in \mathbb{R}^{\bar{m}}$ is control input. The matrix $\mathbf{M}$ consists of inertia like terms, matrix $\mathbf{C}$ contains the Coriolis like terms, $\mathbf{G}$ represents terms dependent only on $\mathbf{q}$ and $\mathbf{B}$ is the input matrix representing the manner in which the control input influence the coordinates.

Definition 1.1 ([42, 43]) A VHC of $\bar{k}$th order for the system (1.9) is defined as $\mathbf{h}(\mathbf{q}) = 0$, where $\mathbf{h} : \mathbb{R}^{\bar{n}} \mapsto \mathbb{R}^{\bar{k}}$ is smooth, $\text{rank}(d\mathbf{h}_\mathbf{q}) = \bar{k} \forall \mathbf{q} \in \mathbf{h}^{-1}(0)$ and the set

$$\Gamma = \{(\mathbf{q}, \dot{\mathbf{q}}) : \mathbf{h}(\mathbf{q}) = 0, d\mathbf{h}_\mathbf{q}\dot{\mathbf{q}} = 0\}$$

1.4 Virtual Holonomic Constraint

is control invariant, i.e. there exists of some smooth control $\tau(\mathbf{q}, \dot{\mathbf{q}})$, such that Γ is positively invariant for the closed-loop system.

Here, the control input $\tau(\mathbf{q}, \dot{\mathbf{q}})$ enforces the VHC $\mathbf{h}(\mathbf{q}) = 0$ and the associated set Γ is called the *constraint manifold*. Further, the VHC $\mathbf{h}(\mathbf{q}) = 0$ is defined as a *regular VHC* if the output function $\mathbf{e} = \mathbf{h}(\mathbf{q})$ yields a relative degree 2.

For an Euler–Lagrangian system with underactuation 1, i.e. $\mathbf{q} \in \mathbb{R}^n$ and $\tau \in \mathbb{R}^{n-1}$, a stabilizing VHC can be expressed as

$$q_1 = \phi_1(q_n),$$
$$q_2 = \phi_2(q_n),$$
$$\vdots$$
$$q_{n-1} = \phi_{n-1}(q_n).$$

The VHCs can be expressed in a vector form as

$$\mathbf{q} = \hat{\boldsymbol{\phi}}(q_n) = \left[\boldsymbol{\phi}(q_n)^T \; q_n\right]^T, \tag{1.10}$$

where $\boldsymbol{\phi}(q_n) = [\phi_1(q_n) \; \phi_2(q_n) \; \ldots \; \phi_{n-1}(q_n)]^T$. The VHC defined in (1.10) is a feasible constraint manifold in accordance to Definition 1.1, if the constraint can be made invariant for a suitable choice of control law $\tau(\mathbf{q}, \dot{\mathbf{q}})$ [43, 44].

Lemma 1.1 *A VHC* $\mathbf{q} = \hat{\boldsymbol{\phi}}(q_n)$ *is feasible for all* $q_n \in S^1$ *if*

$$Im[\mathbf{M}(\hat{\boldsymbol{\phi}}(q_n))\hat{\boldsymbol{\phi}}'(q_n)] \cap \mathbf{B}(\hat{\boldsymbol{\phi}}(q_n)) = \{0\}.$$

Using Lemma 1.1, it can be shown that every regular VHC $\{\mathbf{q} = \hat{\boldsymbol{\phi}}(q_n) \; \forall \; q_n \in S^1\}$ is a feasible [42]. Subsequently, Lemma 1.1 can be written as

$$\mathbf{B}^{\perp}(\hat{\boldsymbol{\phi}}(q_n))\mathbf{M}(\hat{\boldsymbol{\phi}}(q_n))\hat{\boldsymbol{\phi}}'(q_n) \neq 0 \; \forall \; q_n \in S^1. \tag{1.11}$$

The condition in (1.11) provides a scheme to verify the feasibility of a VHC for an Euler–Lagrange system with 1 underactuation.

The error between the actual coordinates and the VHC can be defined as

$$\mathbf{e} = \bar{\mathbf{q}} - \boldsymbol{\phi}(q_n), \tag{1.12}$$

where $\bar{\mathbf{q}} = [q_1 \; q_2 \; \ldots \; q_{n-1}]$. The error system can be derived by differentiating (1.12) twice with respect to time as

$$\ddot{\mathbf{e}} = \ddot{\bar{\mathbf{q}}} - \boldsymbol{\phi}'(q_n)\ddot{q}_n - \boldsymbol{\phi}''(q_n)\dot{q}_n^2,$$
$$= \begin{bmatrix} \mathbf{I}_{n-1} & -\boldsymbol{\phi}'(q_n) \end{bmatrix} \begin{bmatrix} \ddot{\bar{\mathbf{q}}} \\ \ddot{q}_n \end{bmatrix} - \boldsymbol{\phi}''(q_n)\dot{q}_n^2,$$
$$= \begin{bmatrix} \mathbf{I}_{n-1} & -\boldsymbol{\phi}'(q_n) \end{bmatrix} \ddot{\mathbf{q}} - \boldsymbol{\phi}''(q_n)\dot{q}_n^2. \tag{1.13}$$

Utilizing the system dynamics (1.9) in (1.13), one can obtain

$$\ddot{\mathbf{e}} = \begin{bmatrix} \mathbf{I}_{n-1} & -\boldsymbol{\phi}'(q_n) \end{bmatrix} \mathbf{M}^{-1}(\mathbf{q}) \{-\mathbf{C}(\mathbf{q},\dot{\mathbf{q}})\dot{\mathbf{q}} - \mathbf{G}(\mathbf{q}) + \mathbf{B}(\mathbf{q})\boldsymbol{\tau}\} - \boldsymbol{\phi}''(q_n)\dot{q}_n^2,$$
$$\Rightarrow \ddot{\mathbf{e}} = \bar{\mathbf{M}}(\mathbf{q}, \boldsymbol{\phi}'(q_n))\{-\mathbf{C}(\mathbf{q},\dot{\mathbf{q}})\dot{\mathbf{q}} - \mathbf{G}(\mathbf{q})\} - \boldsymbol{\phi}''(q_n)\dot{q}_n^2 + \bar{\mathbf{B}}(\mathbf{q}, \boldsymbol{\phi}'(q_n))\boldsymbol{\tau}, \tag{1.14}$$

where $\bar{\mathbf{M}}(\mathbf{q}, \boldsymbol{\phi}'(q_n)) = \begin{bmatrix} \mathbf{I}_{n-1} & -\boldsymbol{\phi}'(q_n) \end{bmatrix} \mathbf{M}^{-1}(\mathbf{q})$ and $\bar{\mathbf{B}}(\mathbf{q}, \boldsymbol{\phi}'(q_n)) = \bar{\mathbf{M}}(\mathbf{q}, \boldsymbol{\phi}'(q_n)) \mathbf{B}(\mathbf{q})$. The constraint manifold Γ is locally exponentially stabilizable by the control law

$$\boldsymbol{\tau} = \bar{\mathbf{B}}^{-1}(\mathbf{q}, \boldsymbol{\phi}'(q_n))\left\{\bar{\mathbf{M}}(\mathbf{q}, \boldsymbol{\phi}'(q_n))\{\mathbf{C}(\mathbf{q},\dot{\mathbf{q}})\dot{\mathbf{q}} + \mathbf{G}(\mathbf{q})\} + \boldsymbol{\phi}''(q_n)\dot{q}_n^2 - k_v\dot{\mathbf{e}} - k_p\mathbf{e}\right\}, \tag{1.15}$$

which results in the closed loop error dynamics written as

$$\ddot{\mathbf{e}} + k_v\dot{\mathbf{e}} + k_p\mathbf{e} = 0. \tag{1.16}$$

The feasibility condition in Lemma 1.1 can be rewritten as

$$\mathbf{B}^{\perp}(\hat{\boldsymbol{\phi}}(q_n))\mathbf{M}(\hat{\boldsymbol{\phi}}(q_n))\hat{\boldsymbol{\phi}}'(q_n) = \delta(q_n), \tag{1.17}$$

where $\delta(q_n)$ is a continuous function bounded away from zero. Instead of guessing possible holonomic constraints and validating their feasibility, one can utilize this condition to generate feasible VHCs by solving the resulting ODE. By choosing a set of odd functions for $\phi_2(q_n), \phi_3(q_n), \ldots, \phi_{n-1}(q_n)$, with the same time period as q_n, i.e.

$$\phi_j(-q_n) = -\phi_j(q_n) \text{ for } j = 2, \ldots, n-1,$$

one can obtain the remaining virtual constraint satisfying the condition (1.17) as

$$\begin{bmatrix} b_1(\phi_1, q_n) & \ldots & b_{n-1}(\phi_1, q_n) & b_n(\phi_1, q_n) \end{bmatrix} \begin{bmatrix} \phi_1' & \ldots & \phi_{n-1}'(q_n) & 1 \end{bmatrix}^T = \delta(q_n), \tag{1.18}$$

where

$$\begin{bmatrix} b_1(\phi_1, q_n) & \ldots & b_{n-1}(\phi_1, q_n) & b_n(\phi_1, q_n) \end{bmatrix} = \mathbf{B}^{\perp}(\mathbf{q})\mathbf{M}(\mathbf{q})\Big|_{\mathbf{q} = [\phi_1 \ \phi_2(q_n) \ \ldots \ \phi_{n-1}(q_n) \ q_n]}.$$

1.4 Virtual Holonomic Constraint

Hence, the condition (1.18) can be expressed as

$$\phi'_1 = \frac{1}{b_1(\phi_1, q_n)} \left\{ \delta(q_n) - \sum_{i=2}^{n-1} b_i(\phi_1, q_n)\phi'_j(q_n) - b_n(\phi_1, q_n) \right\}. \tag{1.19}$$

Solving (1.19) would yield ϕ_1 that results in a feasible VHC. For this reason, (1.19) is known as *VHC Generator*.

The motion of the system on the VHC manifold can be obtained by multiplying the system dynamics (1.9) by $\mathbf{B}^{\perp}(\mathbf{q})$ as

$$\mathbf{B}^{\perp}(\hat{\boldsymbol{\phi}})\mathbf{M}(\hat{\boldsymbol{\phi}})\hat{\boldsymbol{\phi}}'\ddot{q}_n + \mathbf{B}^{\perp}(\hat{\boldsymbol{\phi}})\left\{\mathbf{M}(\hat{\boldsymbol{\phi}})\hat{\boldsymbol{\phi}}''\dot{q}_n^2 + \mathbf{C}((\hat{\boldsymbol{\phi}}), \hat{\boldsymbol{\phi}}'\dot{q}_n)\hat{\boldsymbol{\phi}}'\dot{q}_n + \mathbf{G}(\hat{\boldsymbol{\phi}})\right\} = 0. \tag{1.20}$$

The constraint dynamics (1.20) can be rewritten using (1.17) as

$$\ddot{q}_n = -\frac{\mathbf{B}^{\perp}(\hat{\boldsymbol{\phi}})}{\delta(q_n)}\left\{\mathbf{M}(\hat{\boldsymbol{\phi}})\hat{\boldsymbol{\phi}}''\dot{q}_n^2 + \mathbf{C}((\hat{\boldsymbol{\phi}}), \hat{\boldsymbol{\phi}}'\dot{q}_n)\hat{\boldsymbol{\phi}}'\dot{q}_n + \mathbf{G}(\hat{\boldsymbol{\phi}})\right\}. \tag{1.21}$$

The system dynamics on the constraint manifold (1.21) describes the behavior of the unactuated system coordinate q_n subject to the enforcement of the virtual constraint $\boldsymbol{\phi}$.

For a better understanding of the VHC-based approach, the feasibility of VHC for a double pendulum system and the design of the stabilizing control law have been discussed next.

1.4.2 Virtual Holonomic Constraint for Double Pendulum

A double pendulum system has been considered here to demonstrate the utilization of VHCs to achieve a desired motion. The state vector are and The matrices of the dynamic equation (1.9) for the double pendulum system illustrated in Fig. 1.2 considering equal-unitary mass and length with state vector $\mathbf{q} = [\theta_1 \ \theta_2]$ can be written as

$$\mathbf{M}(\mathbf{q}) = \begin{bmatrix} 2 & \cos(\theta_1 - \theta_2) \\ \cos(\theta_1 - \theta_2) & 1 \end{bmatrix}, \mathbf{C}(\mathbf{q}, \dot{\mathbf{q}}) = \begin{bmatrix} 0 & \sin(\theta_1 - \theta_2)\dot{\theta}_2 \\ \sin(\theta_1 - \theta_2)\dot{\theta}_1 & 0 \end{bmatrix},$$
$$\mathbf{G}(\mathbf{q}) = \begin{bmatrix} -2g\sin(\theta_1) \\ -g\sin(\theta_2) \end{bmatrix}, \mathbf{B} = \begin{bmatrix} 0 \\ 1 \end{bmatrix}. \tag{1.22}$$

The virtual holonomic constraint for this system is chosen as

$$\theta_1 - \sin(2\theta_2)/4 = 0. \tag{1.23}$$

Fig. 1.2 Schematic diagram of double pendulum

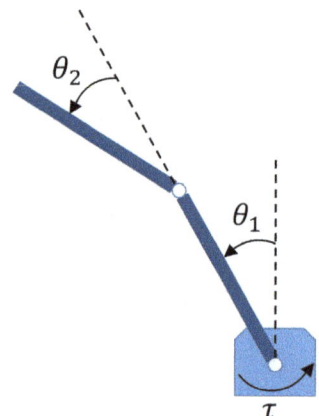

The error in the implementation of the VHC (1.23) can be thus expressed as

$$e = \theta_1 - \sin(2\theta_2)/4. \tag{1.24}$$

With the input matrix $\mathbf{B} = [1\ 0]^T$ for the system (1.22), the orthogonal space of the input matrix becomes $\mathbf{B}^\perp = [0\ 1]$. Verifying condition (1.11) for the virtual constraint (1.23), one can write

$$= \frac{1}{2}\cos(\theta_1 - \theta_2)\cos(2\theta_2) + 1 \neq 0. \tag{1.25}$$

Using the VHC generator (1.19), other feasible constraints can be generated by choosing a non-zero $\delta(\theta_2)$. The stabilizing control law (1.15) for making the virtual constraint (1.23) invariant can be expressed as

$$\tau = \bar{\mathbf{B}}^{-1}(\mathbf{q}, \boldsymbol{\phi}'(\theta_2))\left\{\bar{\mathbf{M}}(\mathbf{q}, \boldsymbol{\phi}'(\theta_2))\{\mathbf{C}(\mathbf{q},\dot{\mathbf{q}})\dot{\mathbf{q}} + \mathbf{G}(\mathbf{q})\} + \boldsymbol{\phi}''(\theta_2)\dot{\theta}_2^2 - k_v\dot{e} - k_p e\right\}, \tag{1.26}$$

where $\bar{\mathbf{M}}(\mathbf{q}, \boldsymbol{\phi}'(\theta_2)) = [1\ -\boldsymbol{\phi}'(\theta_2)]\mathbf{M}^{-1}(\mathbf{q})$, $\bar{\mathbf{B}}(\mathbf{q}, \boldsymbol{\phi}'(\theta_2)) = \bar{\mathbf{M}}(\mathbf{q}, \boldsymbol{\phi}'(\theta_2))\mathbf{B}$. By choosing feedback gains $k_p = K_d = 5$ and an initial configuration $\mathbf{q}(0) = [\pi/2\ 0]^T$, the error in the VHC converges asymptotically to zero as seen in Fig. 1.3 through the stabilizing control law shown in Fig. 1.4. The spatial configuration for the time-lapse motion of the double pendulum are shown in Fig. 1.5. One can see that the unactuated link settles to the vicinity of its stable equilibrium point $\theta_2 = \pi$, while the actuator attempts to enforce the VHC by moving the actuated link.

1.5 Body-Shape Control

Fig. 1.3 VHC error

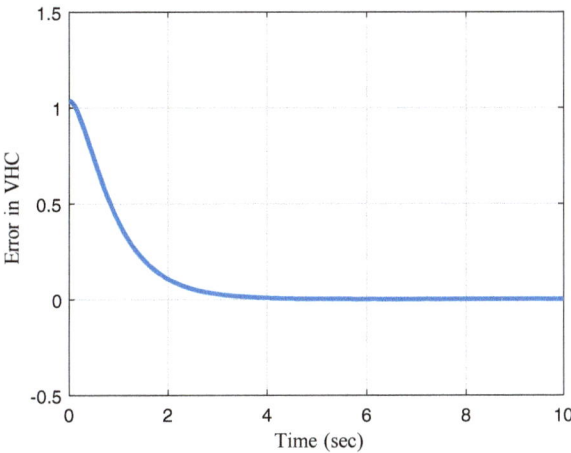

Fig. 1.4 Control input

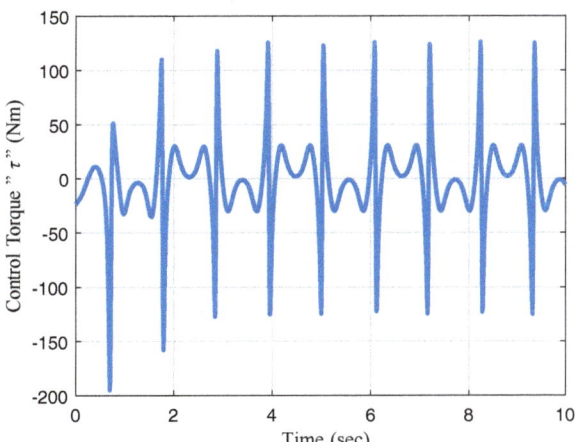

1.5 Body-Shape Control

The generation of controlled undulations is crucial for achieving desired motion for a snake robot. It involves coordinating the joint link actuation to achieve the serpenoid gait function [8, 10] as body shape, known to be responsible for inducing undulations in snakes.

1.5.1 Serpenoid Gait Function

Various species of biological snakes were observed to understand their motion characteristics. These experiments were dedicated toward correlating the surface conditions

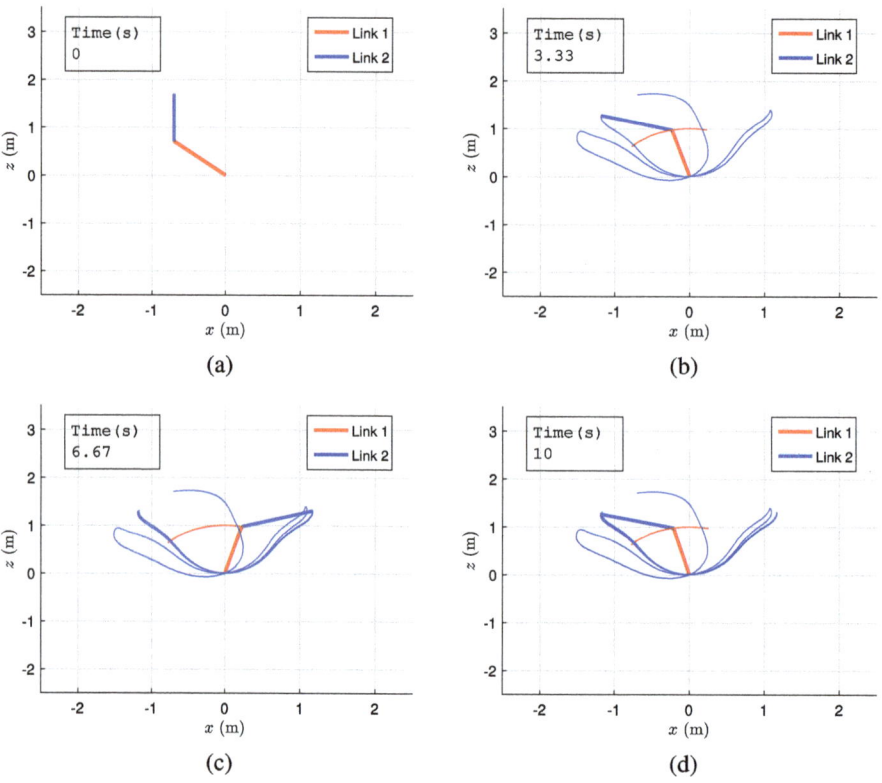

Fig. 1.5 Time lapse (sec) motion of the double pendulum

to the undulation and translation in the snake body. The concept of the serpenoid curve was evolved from the body kinematics of a snake during rectilinear motion. A point equation of a serpenoid curve passing through origin in the $x - y$ plane parameterized by the arc length can be expressed as [12]

$$x(s) = \int_0^s \cos(\phi_\varsigma) d\varsigma, \quad y(s) = \int_0^{ts} \sin(\phi_\varsigma) d\varsigma, \quad \phi_\varsigma := a\cos(b\varsigma) + c\varsigma,$$

where a, b and c are scalar parameters with s being the arc length between the point and the origin. By, varying parameters a, b and c, serpenoid curves with different characteristics can be generated.

This later evolved into an empirical non-continuum gait pattern called *Serpenoid Gait Function* [11] for a robotic snake for it to execute rectilinear motion. A serpenoid gait function can be written as

$$\phi_i = \alpha \sin(\omega t + (i-1)\delta) + \vartheta, \quad i = 1 \ldots n - 1, \tag{1.27}$$

1.5 Body-Shape Control

where $\alpha, \omega, \delta, \vartheta$ are gait parameters that can be varied to generate the required motion. The amplitude α determines the magnitude of transverse undulation of the snake, ω indicates the temporal frequency of the gait function and ϑ is the offset of the gait function. Here, δ is angular lag between consecutive links and its implementation is dependent on the number of links forming the wave. This provides asymmetry to the body shape and results in a non-singular posture ensuring controllability. The velocity of the snake robot and the direction of its motion are sensitive to ω and ϑ, respectively. The frequency of the gait function ω, determines the rate at which the gait wave is propagated along the body of the robot thus controlling the angular velocity of each link which influences the generation of transitional force. The offset ϑ regulates the spatial axis along which the wave propagates, thus moderating the heading of the robot in a global frame.

1.5.2 Virtual Holonomic Constraint-Based Control for Planar Snake Robots

For the planar snake robot, VHC relating the joint angles to the serpenoid gait function is known to be effective [35]. Rewriting the gait function (1.27) as

$$\phi_i = \alpha \sin(\lambda + (i-1)\delta). \tag{1.28}$$

The term ωt is now represented by the variable λ. This line of representation has been done for the reason of velocity control once the desired body shape of the robot is achieved. To achieve the desired body shape, the relative joint angles are required to track the gait function as

$$\theta_i - \theta_{i+1} = \phi_i, \qquad i = 1, \ldots, n-2,$$
$$\theta_{n-1} - \theta_n = \phi_{n-1} + \phi_0.$$

Once the robot assumes this body shape, it propagates according to the frequency and offset of the gait. We note that in most cases, the snake robot is a redundant system and (1.8) is not invertible. Hence, the engineering approach is to update the gait parameters λ and ϕ_0 for tracking a particular velocity and heading-angle through compensators [35] as

$$\ddot{\lambda} = u_\lambda, \quad \ddot{\phi}_0 = u_{\phi_0},$$

where u_λ and u_{ϕ_0} are the pseudo control inputs. A VHC for achieving the serpenoid gait for the snake robot (1.8) can be written [35] as

$$\mathbf{h}(\lambda, \phi_0, \boldsymbol{\theta}) = \mathbf{D}\boldsymbol{\theta} - \boldsymbol{\Phi}(\lambda) - \mathbf{b}_1 \phi_0, \tag{1.29}$$

where, $\boldsymbol{\Phi}(\lambda)$ represents the gait function for all the joints in a vector form. The VHC system dynamics can be obtained by double differentiating (1.29) and substituting robot dynamics (1.8a) to eliminate $\ddot{\boldsymbol{\theta}}$ to yield

$$\ddot{\mathbf{h}} = \mathbf{DM}^{-1}\left(\mathbf{W}\dot{\boldsymbol{\theta}}^2 + l\mathbf{SC}_{\theta}^T \mathbf{f}_R(\boldsymbol{\theta}, \dot{\boldsymbol{\theta}}, \mathbf{p})\right) - \boldsymbol{\Phi}''(\lambda)\dot{\lambda}^2 - \boldsymbol{\Phi}'(\lambda)\ddot{\lambda} - \mathbf{b}_1\ddot{\phi}_0 + \mathbf{DM}^{-1}\mathbf{D}^T\boldsymbol{\tau}. \quad (1.30)$$

The VHC dynamics (1.30) can now be employed to design a suitable control law $\boldsymbol{\tau}$ to enforce a predefined gait function $\boldsymbol{\Phi}(\lambda)$ on the robot joints. A feedback controller presented in [35] to achieve exponential stability of the VHCs can be expressed as

$$\boldsymbol{\tau} = (\mathbf{DM}^{-1}\mathbf{D})^{-1}(-\mathbf{DM}^{-1}\mathbf{W}\dot{\boldsymbol{\theta}}^2 - l\mathbf{DM}^{-1}\mathbf{SC}_{\theta}^T \mathbf{f}_R + \boldsymbol{\Phi}''(\lambda)\dot{\lambda}^2 + \boldsymbol{\Phi}'(\lambda)u_\lambda + \mathbf{b}_1\ddot{\phi}_0$$
$$- \mathbf{K}_P(\mathbf{D}\boldsymbol{\theta} - \boldsymbol{\Phi}(\lambda) - \mathbf{b}_1\phi_0) - \mathbf{K}_D(\mathbf{D}\dot{\boldsymbol{\theta}} - \boldsymbol{\Phi}'(\lambda)\dot{\lambda} - \mathbf{b}_1\dot{\phi}_0)). \quad (1.31)$$

For a deterministic system, where all the system parameters are completely known and the system states $\mathbf{q}$ are measurable, the control law (1.31) results in a stable closed loop system as

$$\ddot{\mathbf{h}} = \mathbf{K}_D\dot{\mathbf{h}} - \mathbf{K}_P\mathbf{h}, \quad (1.32)$$

where $\mathbf{K}_P$ and $\mathbf{K}_D$ can be appropriately chosen to ensure the convergence of the VHC to origin as

$$\mathbf{h} \to 0 \ for \ t \to \infty.$$

1.6 Output-Based Control

Now the reduced system on the constraint manifold is called the constraint system corresponding to the VHC (1.29). The constraint system will yield the values of the dynamic gait parameter λ and ϕ_0 and their higher order derivatives from the velocity and head-angle tracking which will in turn be used in the feedback control (1.31).

1.6.1 Constraint System

The constraint system on the VHC manifold can be expressed as [35, 43],

$$\ddot{\theta}_n = \psi_1 + \psi_2 u_\lambda + \psi_3 u_{\phi_0}, \quad (1.33a)$$
$$\ddot{\mathbf{p}} = \boldsymbol{\Psi}_4\dot{\mathbf{p}} + \boldsymbol{\psi}_5\dot{\theta}_n + \boldsymbol{\psi}_6\dot{\lambda} + \boldsymbol{\psi}_7\dot{\phi}_0 \quad (1.33b)$$

1.6 Output-Based Control

where

$$\psi_1 = -\frac{\mathbf{e}^T\mathbf{MH}\boldsymbol{\Phi}''(\lambda)\dot{\lambda}^2}{\mathbf{e}^T\mathbf{Me}} - \frac{1}{\mathbf{e}^T\mathbf{Me}}\mathbf{e}^T\left(\mathbf{W}\dot{\boldsymbol{\theta}}^2 - l\mathbf{SC}_\theta^T\mathbf{f}_R\right), \quad (1.34a)$$

$$\psi_2 = -\frac{\mathbf{e}^T\mathbf{MH}\boldsymbol{\Phi}'(\lambda)}{\mathbf{e}^T\mathbf{Me}}, \quad (1.34b)$$

$$\psi_3 = -\frac{\mathbf{e}^T\mathbf{MHb}_1}{\mathbf{e}^T\mathbf{Me}}, \quad (1.34c)$$

$$\boldsymbol{\Psi}_4 = -\frac{1}{nm}\mathbf{E}^T\mathbf{Q}_\theta\mathbf{E}, \quad (1.34d)$$

$$\boldsymbol{\psi}_5 = -\frac{1}{nm}\mathbf{E}^T\mathbf{Q}_\theta\mathbf{SC}_\theta\mathbf{e}, \quad (1.34e)$$

$$\boldsymbol{\psi}_6 = -\frac{1}{nm}\mathbf{E}^T\mathbf{Q}_\theta\mathbf{SC}_\theta\mathbf{H}\boldsymbol{\Phi}'(\lambda), \quad (1.34f)$$

$$\boldsymbol{\psi}_7 = -\frac{1}{nm}\mathbf{E}^T\mathbf{Q}_\theta\mathbf{SC}_\theta\mathbf{Hb}_1. \quad (1.34g)$$

The velocity of the CM of the snake robot can be transformed from the global reference frame to the body frame of the head link and can be expressed as

$$\begin{bmatrix} v_t & v_n \end{bmatrix}^T = \begin{bmatrix} \mathbf{u}_{\theta_n} & \mathbf{v}_{\theta_n} \end{bmatrix}^T \dot{\mathbf{p}}, \quad (1.35)$$

where

$$\mathbf{u}_{\theta_n} = \begin{bmatrix} \cos\theta_n & \sin\theta_n \end{bmatrix}, \quad \mathbf{v}_{\theta_n} = \begin{bmatrix} -\sin\theta_n & \cos\theta_n \end{bmatrix}.$$

The body frame velocity dynamics on the constraint manifold can be expressed as

$$\dot{v}_t = f_2(v_t, v_n, \theta_n, \dot{\theta}_n, \lambda, \dot{\lambda}, \phi_0, \dot{\phi}_0), \quad (1.36)$$
$$\dot{v}_n = f_3(v_t, v_n, \theta_n, \dot{\theta}_n, \lambda, \dot{\lambda}, \phi_0, \dot{\phi}_0), \quad (1.37)$$

where

$$f_2 = \mathbf{u}_{\theta_n}^T\boldsymbol{\Psi}_4\mathbf{u}_{\theta_n}v_t + \mathbf{u}_{\theta_n}^T\boldsymbol{\Psi}_4\mathbf{v}_{\theta_n}v_n + \dot{\theta}_n v_n + \mathbf{u}_{\theta_n}^T\boldsymbol{\psi}_5\dot{\theta}_n + \mathbf{u}_{\theta_n}^T\boldsymbol{\psi}_6\dot{\lambda} + \mathbf{u}_{\theta_n}^T\boldsymbol{\psi}_7\dot{\phi}_0,$$
$$f_3 = \mathbf{v}_{\theta_n}^T\boldsymbol{\Psi}_4\mathbf{u}_{\theta_n}v_t + \mathbf{v}_{\theta_n}^T\boldsymbol{\Psi}_4\mathbf{v}_{\theta_n}v_n - \dot{\theta}_n v_t + \mathbf{v}_{\theta_n}^T\boldsymbol{\psi}_5\dot{\theta}_n + \mathbf{v}_{\theta_n}^T\boldsymbol{\psi}_6\dot{\lambda} + \mathbf{v}_{\theta_n}^T\boldsymbol{\psi}_7\dot{\phi}_0.$$

Considering θ_{ref} to be the reference head-angle and v_{ref} to be the desired tangential velocity, the head-angle and tangential velocity error can be written as

$$\tilde{\theta}_n = \theta_n - \theta_{ref}, \quad \tilde{v}_t = v_t - v_{ref}.$$

The head-angle error and body frame velocity error dynamics along with the dynamic gait parameter compensators can be expressed as

$$\ddot{\tilde{\theta}}_n = (\psi_1 - \ddot{\theta}_{\text{ref}}) + \psi_2 u_\lambda + \psi_3 u_{\phi_0}, \tag{1.38a}$$
$$\dot{\tilde{v}}_t = f_2(\bar{\mathbf{x}}) - \dot{v}_{\text{ref}}, \tag{1.38b}$$
$$\dot{v}_n = f_3(\bar{\mathbf{x}}), \tag{1.38c}$$
$$\ddot{\lambda} = u_\lambda, \tag{1.38d}$$
$$\ddot{\phi}_0 = u_{\phi_0}, \tag{1.38e}$$

where the state vector for the constraint system can be expressed as

$$\bar{\mathbf{x}} = \begin{bmatrix} v_t & v_n & \theta_n & \dot{\theta}_n & \lambda & \dot{\lambda} & \phi_0 & \dot{\phi}_0 \end{bmatrix}^T.$$

The maneuvering control approach [35] presents a gradual scheme by addressing the head-angle tracking first followed by the velocity tracking for the constraint system.

1.6.2 Head-Angle Control

The head-angle tracking problem has been addressed here considering the head link angle dynamics (1.38a) and the compensator for the gait offset (1.38e). The head-angle system can be expressed as

$$\begin{aligned} \ddot{\theta}_n &= f_1(\theta_n, \dot{\theta}_n, \lambda, \dot{\lambda}, \phi_0, \dot{\phi}_0, u_\lambda) + \psi_3 u_{\phi_0}, \\ \ddot{\phi}_0 &= u_{\phi_0}. \end{aligned} \tag{1.39}$$

The feedback control law adopted in [35] is given as

$$u_{\phi_0} = \frac{1}{\psi_3}\left(\frac{1}{\epsilon}(\dot{\tilde{\theta}}_n + k_n \tilde{\theta}_n)\right) - k_1 \phi_0 - k_2 \dot{\phi}_0, \tag{1.40}$$

where ϵ is a small positive constant, k_n, k_1 and k_2 are control gains to be tuned suitably. The closed loop equations can be written in the form of a *singularly perturbed* system as

$$\begin{aligned} \dot{\tilde{\theta}}_n &= \tilde{\omega}_n, \\ \epsilon \dot{\tilde{\omega}}_n &= \epsilon \left\{ \ddot{\theta}_{ref} + g_1(t, \theta_n, \dot{\theta}_n, \phi_0, \dot{\phi}_0) + \psi_3(k_1\phi_0 + k_2\dot{\phi}_0) \right\} - (\tilde{\omega}_n + k_n \tilde{\theta}_n), \end{aligned} \tag{1.41}$$

where $g_1(t, \theta_n, \dot{\theta}_n, \phi_0, \dot{\phi}_0) = f_1(\theta_n, \dot{\theta}_n, \lambda, \dot{\lambda}, \phi_0, \dot{\phi}_0, u_\lambda)$. The influence of u_λ on the system has been eliminated by using *time-scale separation* analysis. The associated boundary layer dynamics is given as

$$\frac{d\hat{y}}{d\tau} = -\hat{y}, \tag{1.42}$$

1.6 Output-Based Control

where $\hat{y} = \tilde{\omega}_n + k_n \tilde{\theta}_n$ and $\tau = t/\epsilon$. Both the reduced order and the boundary layer dynamics can be seen to exponentially converge to the origin, it being the stable equilibrium point. The closed loop dynamics of the states $(\phi_0, \dot{\phi}_0)$ can be written as

$$\ddot{\phi}_0 + k_2 \dot{\phi}_0 + k_1 \phi_0 = \frac{1}{\psi_3} \left\{ \frac{1}{\epsilon} (\dot{\tilde{\theta}}_n + k_n \tilde{\theta}_n) \right\}. \tag{1.43}$$

The right-hand side of (1.43) can be deduced to be bounded from the solution of (1.42). This ensures the convergence of (1.43) to its global stable equilibrium, the origin with appropriate choices of gain k_1 and k_2.

1.6.3 Velocity Control

To track the tangential velocity v_t to a reference velocity and bounding the normal velocity v_n about zero, the velocity error dynamics can be expressed as

$$\dot{\tilde{v}}_t = f_2(\theta_n, \dot{\theta}_n, \lambda, \phi_0, \dot{\phi}_0, \tilde{v}_t, v_n) + \mathbf{u}_{\theta_n}^T \boldsymbol{\psi}_6 \dot{\lambda} - (dv_{ref})_p \dot{p}, \tag{1.44a}$$

$$\dot{v}_n = f_2(\theta_n, \dot{\theta}_n, \lambda, \dot{\lambda}, \phi_0, \dot{\phi}_0, \tilde{v}_t, v_n) + \mathbf{v}_{\theta_n}^T \boldsymbol{\Psi}_4 \mathbf{v}_{\theta_n} v_n, \tag{1.44b}$$

$$\ddot{\lambda} = u_\lambda. \tag{1.44c}$$

The input–output linearization-based control law subject to the output function, $z = \dot{\lambda} + K_\lambda \tilde{v}_t$ is given as [35]

$$u_\lambda = -K_z(\dot{\lambda} + K_\lambda \tilde{v}_t) - K_\lambda \left(f_2(\cdot) + \mathbf{u}_{\theta_n}^T \boldsymbol{\psi}_6 \dot{\lambda} - (dv_{ref})_p \dot{p} \right). \tag{1.45}$$

The performance of this control approach has been adopted as the benchmark, to investigate the efficiency of the various robust control techniques to be proposed in the upcoming chapters.

1.7 Toward a Practical Control Framework

The state-of-the-art control approach described theretofore, is a complicated framework for implementation and might encounter some serious issues when being deployed in a practical situation. This section provides a detailed assessment of such issues and details how the works described in subsequent chapters are a step toward bridging these research gaps.

1.7.1 Robustness for Planar Snake Robots

The control approaches presented in Sects. 1.5 and 1.6 have been proposed for a deterministic model of a planar snake robot [35]. It implies that the controller mentioned above requires exact knowledge of the system to ensure stability and efficient tracking. The controller proposed are methods to track as well as evolve the serpenoid gait parameters automatically. For situations where the a priori knowledge of the system becomes a conservative assumption, the aforementioned approach cannot guarantee stable tracking performance. Though achieving desired motion even in structured surroundings is not easy, uncertainties in the environment pose additional challenges to the control law design problem. Notably, friction coefficients between surfaces in contact are notoriously uncertain and can deteriorate the tracking performance. Since this has to be overcome for field implementation, robust control techniques effective in solving the tracking problem of a planar snake robot with realistic uncertainties in the system are needed.

SMC approach is a control technique employed to achieve robustness toward bounded uncertainties with known constant upper bound. ASMC requires the uncertainties to be bounded but does not need the values of the bound to be known a priori. Adaptation law of the ASMC alleviates the overestimation problem of the switching gain as well. Though the ASMC methodology doesn't need the uncertainty bounds to be known, nevertheless, the uncertainties are required to be bounded. Moreover, these approaches result in chattering which is undesired for any practical application. Addressing these concerns, a TDC approach can be adopted which utilizes the input–output data from the previous instant to estimate the uncertainty in the system. This methodology requires the system uncertainties to be slowly varying that results into a bounded time-delayed estimation error. An adaptive-robust control algorithm is utilized in addition to the TDC law, to alleviate the effect of the estimation error on the tracking accuracy, thus referred to as ARTDC.

1.7.2 Multi-layered Control Methodology

The planar snake robot model presented in Sect. 1.2 [20, 21] involves the dynamic equations of motion for the angular motions of the individual links θ along with the translation of the CM of the robot **p**. The control torque input to each joint τ can be seen to be mapped to the angular dynamics (1.8a). The angular and linear velocities generated, influence the motion of the robot CM. Hence, the conventional approach partitions the whole dynamics into the body-shape system realized by the torque inputs and the output system that remain unactuated. Subsequently, a multi-layer approach is adopted to solve the tracking problem by addressing the partitioned systems one after another. As the outputs of the system are not mapped to the control input, pseudo-inputs are chosen which influence the computation of the control signals to the actuators for achieving tracking. To circumvent this complicated approach, direct mapping between the output space dynamics and the control input is required.

Differential flatness is a method to transform a system into an output system of same dimension as that of the inputs to simplify trajectory planning and control design. A set of variables known as *flat outputs* are chosen so that the states and the inputs can be recreated from the output and its finite number of higher order derivatives. Thereafter, the original system can be transformed into the *flat system* over which controller design becomes much more convenient. This methodology has been exploited to establish a mapping between the output and the input space of a snake robot.

1.7.3 Modeling Other Modes of Propagation

As has been already highlighted, the motion of a snake robot on a flat plane has been extensively studied [10, 11, 20, 21]. Although to exploit the true potential of a system like robotic snake, it is essential to study the other aspects of snake robot motion as well. This calls for a detailed study on the other modes of propagation which would be a step further in the field of serpentine robotics.

Among the other modes, the motion of a snake robot inside a constrained space is particularly interesting and has a range of prospective applications. This includes inspection of industrial pipes and channels, surveillance, disaster relief, etc. There is a high probability that the robot might not find ground contact and have to move by other means. Hence, to deploy a snake robot in such constrained or tight spaces, the motion of the robot inside such a space is required to be studied in detail. A primary step towards the same is to model a 2D representation of a pipe facilitating motion for a robotic snake without friction from the under-surface. This would enable a robotic snake to generate propagating force from whichever surface it is in contact with and achieve the desired motion.

1.8 The Theme

Drawing inspiration from the above discussion, the theme of this book is framed as
to design adaptive robust control strategies to address uncertainties and disturbances in planar snake robot while alleviating under- and overestimation of gains and circumventing conservative assumptions.

In view of this theme, the major segments of the book can be summarized as

- Uncertainties in the snake robot dynamics have been considered to be bounded and a robust sliding-mode control law based on an appropriate choice of sliding

surface has been designed. Also, an adaptation law to ease the overestimation of switching gain has been discussed thus reducing the required magnitude of control input.
- An artificial TDC approach has been presented to handle unbounded slowly varying uncertainties in the system.
- An ARTDC methodology has been detailed where the time-delayed estimation is employed with a feedback and a switching control law in addition to an adaptation law for the switching gain to alleviate the effect of estimation error on the tracking performance.
- To establish a direct mapping between the output space and the input space for the ease of trajectory planning and control design, a differential-flatness-based control approach has been discussed as well.
- To propose a mathematical model for the motion of a planar snake robot inside a uniform pipe or channel that aids toward the study of constraint motion of a snake robot as well as assist in controller design.

1.9 Organization of the Book

The organization of this book is detailed as follows:

- Chapter 2 presents a SMC law for the velocity and head-angle tracking of a planar snake robot in the presence of bounded uncertainties with known upper bound. With this approach, the robot can operate effectively under uncertain environments, say with variable friction exhibiting efficient tracking performance. Further, an adaptation law to update the switching gain of the control law is shown to alleviate the overestimation problem of the switching gain, thus reducing the magnitude of the required control input. The requirement of knowledge regarding the bounds in parameter uncertainty has also been skirted through the ASMC approach.
- Chapter 3 discusses a dual-layer time-delayed controller for maneuvering control of planar snake robot with unbounded uncertainties. This approach utilizes delay-based estimation strategy for uncertainties in the dynamic model to compute an efficient control law. The boundedness assumption for the system uncertainties can be evaded in this approach as long as their varying slowly is extended. The stability bound for the system states corresponding to the Uniformly Ultimately Bounded (UUB) sense of stability for the closed-loop system, can be expressed as the ratio of the norm of estimation error over the minimum eigenvalue of the gain matrix. Thus, the region of convergence can be reduced by sufficiently increasing the feedback gain matrix.
- Chapter 4 proposes a dual-layer ARTDC for head-angle and velocity control of a planar snake robot with unbounded uncertainties. The stability bound has a detrimental effect on the tracking performance of the TDC methodology. Hence, a switching control law along with a novel adaptation law has been employed to get an enhanced handle on the stability. The adaptation law utilizes a dual-rate

algorithm to alleviate overestimation–underestimation problem of the switching gain achieving improved tracking performance.

- Chapter 5 explores a differential-flatness-based methodology toward a simplified control approach obviating the partitioned framework adopted in literature owing to the intricate dynamics of a snake robot. In the absence of a direct relationship between the output space and the control inputs, flatness has been proposed to establish a mapping between the output trajectory and the control torques. This makes trajectory specification and control design convenient and efficient.
- Chapter 6 Motion of a snake robot inside a constrained space has been modeled to obtain the dynamic equations. An elliptical link profile has been considered for the robot segment to detect contact with the walls of the channel and estimate contact forces. A Hertz contact model has been utilized to compute the contact force, whereas the traction force has been obtained through Coulomb friction model. Moments due to these forces have been computed by cross-products. All these components have been combined to obtain the dynamical model following which, the control approaches presented above have been employed on the proposed model to investigate tracking efficiency.

References

1. Gray, J., Lissmann, H.W.: The kinetics of locomotion of the grass-snake. J. Exp. Biol. **26**(4), 354–367 (1950). http://jeb.biologists.org/content/26/4/354
2. Lissmann, H.W.: Rectilinear locomotion in a snake (boa occidentalis). J. Exp. Biol. **26**(4), 368–379 (1950). http://jeb.biologists.org/content/26/4/368
3. Jayne, B.C.: Kinematics of terrestrial snake locomotion. Copeia pp. 915–927 (1986)
4. Jayne, B.C.: Muscular mechanisms of snake locomotion: an electromyographic study of the sidewinding and concertina modes of crotalus cerastes, nerodia fasciata and elaphe obsoleta. J. Exp. Biol. **140**(1), 1–33 (1988)
5. Jayne, B.C., Davis, J.D.: Kinematics and performance capacity for the concertina locomotion of a snake (coluber constrictor). J. Exp. Biol. **156**(1), 539–556 (1991). http://jeb.biologists.org/content/156/1/539
6. Secor, S.M., Jayne, B.C., Bennett, A.F.: Locomotor performance and energetic cost of sidewinding by the snake crotalus cerastes. J. Exp. Biol. **163**(1), 1–14 (1992)
7. Chen, J., Friesen, W., Iwasaki, T.: Mechanisms underlying rhythmic locomotion: body-fluid interaction in undulatory swimming. J. Exp. Biol. **214**(4), 561–574 (2011)
8. Umetani, Y., Hirose, S.: Biomechanical study on serpentine locomotion. Trans. Soc. Inst. Control Eng. **8**(6), 724–731 (1972)
9. Bennet, S., McConnell, T., Trubatch, S.L.: Quantitative analysis of the speed of snakes as a function of peg spacing. J. Exp. Biol. **60**(1), 161–165 (1974). http://jeb.biologists.org/content/60/1/161
10. Hirose, S., Umetani, Y.: Kinematic control of active cord mechanism with tactile sensors. Trans. Soci. Inst. Control Eng. **12**(5), 543–547 (1976)
11. Hirose, S., Morishima, A.: Design and control of a mobile robot with an articulated body. Int. J. Robot. Res. **9**(2), 99–114 (1990). https://doi.org/10.1177/027836499000900208
12. Hirose, S.: Biologically Inspired Robots: Snake-Like Locomotors and Manipulators. Oxford University Press, Oxford (1993)

13. Burdick, J.W., Radford, J., Chirikjian, G.S.: A'sidewinding'locomotion gait for hyper-redundant robots. Adv. Robot. **9**(3), 195–216 (1994)
14. Chirikjian, G.S., Burdick, J.W.: The kinematics of hyper-redundant robot locomotion. IEEE Trans. Robot. Autom. **11**(6), 781–793 (1995)
15. Ostrowski, J., Burdick, J.: The geometric mechanics of undulatory robotic locomotion. Int. J. Robot. Res. **17**(7), 683–701 (1998)
16. Krishnaprasad, P.S., Tsakiris, D.P.: G-snakes: nonholonomic kinematic chains on lie groups. In: Proceedings of 1994 33rd IEEE Conference on Decision and Control, vol. 3, pp. 2955–2960 (1994). https://doi.org/10.1109/CDC.1994.411343
17. Shugen: Analysis of creeping locomotion of a snake-like robot. Adv. Robot.**15**(2), 205–224 (2001). https://doi.org/10.1163/15685530152116236
18. Migadis, G., Kyriakopoulos, K.J.: Design and forward kinematic analysis of a robotic snake. In: Proceedings of International Conference on Robotics and Automation, vol. 4, pp. 3493–3498 (1997). https://doi.org/10.1109/ROBOT.1997.606876
19. Nilsson, M.: Serpentine locomotion on surfaces with uniform friction. In: 2004 IEEE/RSJ International Conference on Intelligent Robots and Systems (IROS) (IEEE Cat. No.04CH37566), vol. 2, pp. 1451–1455 (2004). https://doi.org/10.1109/IROS.2004.1389649
20. Liljeback, P., Pettersen, K.Y., Stavdahl, O.: Modelling and control of obstacle-aided snake robot locomotion based on jam resolution. In: 2009 IEEE International Conference on Robotics and Automation, pp. 3807–3814 (2009). https://doi.org/10.1109/ROBOT.2009.5152273
21. Liljeback, P., Pettersen, K.Y., Stavdahl, C., Gravdahl, J.T.: Controllability and stability analysis of planar snake robot locomotion. IEEE Trans. Autom. Control **56**(6), 1365–1380 (2011). http://orcid.org/10.1109/TAC.2010.2088830
22. Liljebäck, P., Pettersen, K.Y., Stavdahl, Ø., Gravdahl, J.T.: A review on modelling, implementation, and control of snake robots. Robot. Auton. Syst. **60**(1), 29–40 (2012)
23. Saito, M., Fukaya, M., Iwasaki, T.: Modeling, analysis, and synthesis of serpentine locomotion with a multilink robotic snake. IEEE Control Syst. Mag. **22**(1), 64–81 (2002)
24. Sato, M., Fukaya, M., Iwasaki, T.: Serpentine locomotion with robotic snakes. IEEE Control Syst. Mag. **22**(1), 64–81 (2002). http://orcid.org/10.1109/37.980248
25. Liljebäck, P., Pettersen, K.Y., Stavdahl, Ø., Gravdahl, J.T.: Snake robots: modelling, mechatronics, and control. Springer Science & Business Media, Berlin (2012)
26. Sarrigeorgidis, K., Kyriakopoulos, K.J.: Stabilization and trajectory tracking of a robotic snake. In: Proceedings of the 36th IEEE Conference on Decision and Control, vol. 3, pp. 3061–3062 (1997). https://doi.org/10.1109/CDC.1997.657919
27. Paap, K.L., Kirchner, F., Klaassen, B.: Motion control scheme for a snake-like robot. In: Proceedings 1999 IEEE International Symposium on Computational Intelligence in Robotics and Automation. CIRA'99 (Cat. No.99EX375), pp. 59–63 (1999). https://doi.org/10.1109/CIRA.1999.809947
28. Prautsch, P., Mita, T.: Control and analysis of the gait of snake robots. In: Proceedings of the 1999 IEEE International Conference on Control Applications (Cat. No.99CH36328), vol. 1, pp. 502–507 (1999). https://doi.org/10.1109/CCA.1999.806692
29. Date, H., Hoshi, Y., Sampei, M.: Locomotion control of a snake-like robot based on dynamic manipulability. In: Proceedings. 2000 IEEE/RSJ International Conference on Intelligent Robots and Systems (IROS 2000) (Cat. No.00CH37113), vol. 3, pp. 2236–2241 (2000). https://doi.org/10.1109/IROS.2000.895301
30. Date, H., Sampei, M., Nakaura, S.: Control of a snake robot in consideration of constraint force. In: Proceedings of the 2001 IEEE International Conference on Control Applications (CCA'01) (Cat. No.01CH37204), pp. 966–971 (2001). https://doi.org/10.1109/CCA.2001.973995
31. Matsuno, F., Suenaga, K.: Control of redundant snake robot based on kinematic model. In: Proceedings of the 41st SICE Annual Conference. SICE 2002, vol. 3, pp. 1481–1486 (2002). https://doi.org/10.1109/SICE.2002.1196525
32. Yamada, T., Tanaka, K., Yamakita, M.: Winding and task control of snake like robot. SICE 2003 Annual Conference (IEEE Cat. No.03TH8734), vol. 3, pp. 3059–3063 (2003)

References

33. Transeth, A.A., van de Wouw, N., Pavlov, A., Hespanha, J.P., Pettersen, K.Y.: Tracking control for snake robot joints. In: 2007 IEEE/RSJ International Conference on Intelligent Robots and Systems, pp. 3539–3546 (2007). https://doi.org/10.1109/IROS.2007.4399174
34. Ishikawa, M.: Iterative feedback control of snake-like robot based on principal fibre bundle modelling. Int. J. Adv. Mechatron. Syst. **1**(3), 175–182 (2009)
35. Mohammadi, A., Rezapour, E., Maggiore, M., Pettersen, K.Y.: Maneuvering control of planar snake robots using virtual holonomic constraints. IEEE Trans. Control Syst. Technol. **24**(3), 884–899 (2016). http://orcid.org/10.1109/TCST.2015.2467208
36. Zhang, A., Ma, S., Li, B., Wang, M., Guo, X., Wang, Y.: Adaptive controller design for underwater snake robot with unmatched uncertainties. Sci. China Inf. Sci. **59**(5), 052205 (2016). https://doi.org/10.1007/s11432-015-5421-8
37. Ariizumi, R., Takahashi, R., Tanaka, M., Asai, T.: Head-trajectory-tracking control of a snake robot and its robustness under actuator failure. IEEE Trans. Control Syst. Technol. 1–9 (2018). https://doi.org/10.1109/TCST.2018.2866964
38. Wang, G., Yang, W., Shen, Y., Shao, H., Wang, C.: Adaptive path following of underactuated snake robot on unknown and varied frictions ground: Theory and validations. IEEE Robot. Autom. Lett. **3**(4), 4273–4280 (2018). http://orcid.org/10.1109/LRA.2018.2864602
39. Hannigan, E., Song, B., Khandate, G., Haas-Heger, M., Yin, J., Ciocarlie, M.: Automatic snake gait generation using model predictive control (2019)
40. Travers, M., Whitman, J., Choset, H.: Shape-based coordination in locomotion control. Int. J. Robot. Res. **37**(10), 1253–1268 (2018). https://doi.org/10.1177/0278364918761569
41. Sartoretti, G., Paivine, W., Shi, Y., Wu, Y., Choset, H.: Distributed learning of decentralized control policies for articulated mobile robots. IEEE Trans. Robot. **35**(5), 1109–1122 (2019). http://orcid.org/10.1109/TRO.2019.2922493
42. Consolini, L., Maggiore, M.: Virtual holonomic constraints for Euler-Lagrange systems. IFAC Proc. Vol. **43**(14), 1193–1198 (2010). https://doi.org/10.3182/20100901-3-IT-2016.00107. 8th IFAC Symposium on Nonlinear Control Systems
43. Maggiore, M., Consolini, L.: Virtual holonomic constraints for Euler–Lagrange systems. IEEE Trans. Autom. Control **58**(4), 1001–1008 (2013)
44. Consolini, L., Maggiore, M.: Control of a bicycle using virtual holonomic constraints. Automatica **49**(9), 2831 – 2839 (2013). https://doi.org/10.1016/j.automatica.2013.05.021

Chapter 2
Adaptive Sliding-Mode Control for Velocity and Head-Angle Tracking

Abstract The ability of a control scheme for a mobile robot to ensure satisfactory performance in an unstructured or unknown environment is what makes the control law unique. Various sources of uncertainties pose a serious challenge to the tracking performance of the system. The state-of-the-art control law presented in Chap. 1 cannot ensure stability in the presence of uncertainties due to feedback linearization methodology, as it requires exact cancellation of the nonlinearities in the system [1, 2]. Moreover, singular perturbation-based approach is prone to generate high control gains which may jeopardize the integrity of the actuators [1, 2]. This is where *robust control* techniques step in, to assure stable and acceptable performance in the presence of uncertainties. To imitate variation of ground condition, time-varying uncertainties have been induced through the friction coefficients in the planar snake robot model. These uncertainties have been assumed to be bounded with a known upper bound to implement a *Sliding-Mode Control* (SMC) law with the aim of achieving efficient head-angle and velocity tracking. Furthermore, to relax the constraint on the uncertainty bound and also to solve the overestimation of switching gain, an adaptive SMC has been proposed to improve the tracking performance.

This chapter primarily focuses on proposing a sliding-mode-based control law and further extending it with an adaptation law for planar snake robot to achieve efficient velocity and head-angle tracking. An appropriate sliding surface has been designed such that the reduced order system is stable. Further, a reaching and switching control law has been proposed to ensure finite-time reachability of the system states to the sliding manifold in the presence of uncertainties with known bounds [3]. A generic adaptation law has been employed for updating the switching gain to alleviate the overestimation problem of the SMC-based approach. Moreover, the ASMC methodology can be executed without a priori knowledge of the uncertainty bound [4]. Stability analysis using Lyapunov function has been performed for the system utilizing the proposed SMC and ASMC approaches. The analysis with SMC law shows asymptotic stability, whereas, with the ASMC law, UUB stability can be ensured for the closed-loop system. Efficiency of the proposed methods has been verified through simulation studies where the performance of the SMC law has been compared with the state-of-the-art approach presented in Chap. 1, whereas the efficacy of the ASMC law has been compared with the proposed SMC approach.

The rest of the chapter is arranged as follows: The problem is formulated in Sect. 2.1; Sect. 2.3 presents the SMC approach along with the associated results; The ASMC scheme with the adaptation law and the results of the same are discussed in Sect. 2.4; Finally, the chapter is summarized in Sect. 2.6.

2.1 Problem Formulation

The head-angle and translational dynamics on the constrained manifold (1.29) along with the dynamic gait parameter compensators have already been presented in (1.38). These equations have been formulated for a deterministic system where the system dynamics is known. However, uncertainties have been assumed in the friction coefficients thus making the friction force vector unknown. These uncertainties have been incorporated in the reduced order system on the constraint manifold and can be expressed as

$$\ddot{\theta}_n = (\hat{\psi}_1 - \ddot{\theta}_{\text{ref}}) + \psi_2 u_\lambda + \psi_3 u_{\phi_0} + \tilde{\psi}_{\theta_n}, \tag{2.1a}$$

$$\dot{\tilde{v}}_t = \hat{f}_2(\bar{\mathbf{x}}) - \dot{v}_{\text{ref}} + \tilde{\psi}_{v_t}, \tag{2.1b}$$

$$\dot{v}_n = \hat{f}_3(\bar{\mathbf{x}}) + \tilde{\psi}_{v_n}, \tag{2.1c}$$

$$\dot{\lambda} = u_\lambda, \tag{2.1d}$$

$$\ddot{\phi}_0 = u_{\phi_0}, \tag{2.1e}$$

where $\tilde{\psi}_{\theta_n}, \tilde{\psi}_{v_t} = \mathbf{u}_{\theta_n}\tilde{\boldsymbol{\psi}}_p$ and $\tilde{\psi}_{v_n} = \mathbf{v}_{\theta_n}\tilde{\boldsymbol{\psi}}_p$ represent the lumped uncertainties in the head-angle, tangential and normal velocity equations. The nominal parts of ψ_1, f_2 and f_3 are symbolized by $\hat{\psi}_1$, $\hat{f}_2$ and $\hat{f}_3$, respectively. The state vector for the reduced system (2.1a) can be written as

$$\bar{\mathbf{x}} = \begin{bmatrix} \tilde{\theta}_n & \dot{\tilde{\theta}}_n & \tilde{v}_t & v_n & \lambda & \dot{\lambda} & \phi_0 & \dot{\phi}_0 \end{bmatrix}^T.$$

Prior to the utilization of SMC and ASMC frameworks for the aforementioned system (2.1a), these techniques have been introduced and briefed for a generic nonlinear system in the following section.

2.2 Brief Outline of Sliding-Mode and Adaptive Sliding-Mode Control

This section gives a brief account of the conventional SMC and ASMC approaches utilized to acquire robustness toward bounded uncertainties.

2.2.1 Sliding-Mode Control

SMC is a popular robust control technique employed to achieve satisfactory tracking as well as regulation performance in the presence of uncertainties in the system. Assuming a generic time-invariant nonlinear system of the form [1, 5]

$$\dot{\mathbf{x}} = f_1(\mathbf{x}) + f_2(\mathbf{x})\mathbf{u} + \mathbf{d}, \tag{2.2}$$

where $\mathbf{x} \in \mathbb{R}^n$ is the state vector, $\mathbf{u} \in \mathbb{R}^m$ is the input to the system and $\mathbf{d}$ represents all the uncertainties in the system. For such a system, sliding surfaces are chosen as

$$\boldsymbol{\sigma} = g(\mathbf{x}) \in \mathbb{R}^m.$$

A sliding surface [5, 6] is chosen as such, that the reduced order dynamics on the surface should be stable, i.e.

$$\boldsymbol{\sigma} = g(\mathbf{x}) = 0 \;\; \mapsto \;\; \mathbf{x} = 0 \; \text{for} \; t \to \infty.$$

Such a choice of sliding surface would assure that the system stays on the sliding surface once it reaches the same. Now to force the system on the sliding surface, appropriate control law is designed for the sliding system expressed as

$$\dot{\boldsymbol{\sigma}} = g'(\mathbf{x})\dot{\mathbf{x}} = g'(\mathbf{x})\Big(f_1(\mathbf{x}) + f_2(\mathbf{x})\mathbf{u} + \mathbf{d}\Big),$$
$$\Rightarrow \dot{\boldsymbol{\sigma}} = \bar{f}_1(\mathbf{x}) + \bar{f}_2(\mathbf{x})\mathbf{u} + \bar{\mathbf{d}}. \tag{2.3}$$

The control law here is segregated in two parts: *Equivalent Control* and *Switching Control*. The equivalent control primarily provides the control input in the absence of any uncertainty, i.e. to achieve desired performance for the nominal system. The equivalent control can be expressed as [5, 6]

$$\mathbf{u}_{eq} = \bar{f}_2(\mathbf{x})^{-1}\Big(-\bar{f}_1(\mathbf{x}) - \mathbf{K}\boldsymbol{\sigma} \Big), \tag{2.4}$$

whereas the switching control tries to eradicate the effect of uncertainties on the system dynamics. It is called switching control as it executes switching the control input depending upon the direction of the sliding system thus forcing the states back toward the sliding manifold. The switching control law can be written as [5, 6]

$$\mathbf{u}_{sw} = -\eta \bar{f}_2(\mathbf{x})^{-1}\text{sign}(\boldsymbol{\sigma}). \tag{2.5}$$

The cumulative control effort is a combination of the equivalent and switching control input as

$$\mathbf{u} = \mathbf{u}_{eq} + \mathbf{u}_{sw}. \tag{2.6}$$

Upon the application of the control law (2.6) to the system dynamics (2.3), the closed-loop system is obtained as

$$\dot{\boldsymbol{\sigma}} = -\mathbf{K}\boldsymbol{\sigma} - \eta \text{sign}(\boldsymbol{\sigma}) + \bar{\mathbf{d}}. \tag{2.7}$$

Upon further investigation, it can be established that the closed-loop system (2.7) is stable at the origin for any $\eta \geq \|d\|$. This is the primary reason behind the basic assumption for applying SMC that the uncertainties in the system should be bounded and the upper bound of the same should be known. If this assumption holds, the designer can always choose a suitable η that ensures system stability.

2.2.2 Adaptive Sliding-Mode Control

To ensure asymptotic stability, the switching gain η is required to be more than the uncertainty at any given time instant. Due to the absence of any instantaneous information regarding the parameter variation, the switching gain is chosen such that it is greater than the upper bound of the uncertainties. That way, we can ensure the negative definiteness of the derivative of Lyapunov function. This *overestimation* [7, 8] of the switching gain is one of the key issue with SMC-based approach which results in high control input as well as chattering. Also, the availability of the uncertainty bound is not a realistic scenario. This is the reason behind the development of ASMC where the switching gain η is updated according to the system states. A generalized adaptation law can be written as [7, 8]

$$\dot{\eta} = f_\eta(\boldsymbol{\sigma}). \tag{2.8}$$

The initial condition of the switching gain $\eta(0)$ should be a positive controller parameter to be chosen appropriately. The key characteristics of the adaptation law (2.8) can be

- The gradient $\dot{\eta}$ should increase if the sliding variable $\boldsymbol{\sigma}$ also increase and vice versa.
- The switching gain should be a non-negative value for all time.
- The switching gain η should be bounded

For such an adaptation law, the establishment of *ideal sliding mode* is not possible in a real application. What we can achieve is a *real sliding mode* where achieving the sliding mode means bounding the states about the sliding manifold. The introduction of this adaptation law may escalate the computational complexity, but would surely reduce the control input exhibiting improved performance.

With both SMC and ASMC been briefed with introductory examples, the following section illustrates the choice of a suitable sliding function and the SMC-based control law specific to the planar snake robot.

2.3 Sliding-Mode Control for Planar Snake Robot

The velocity and head-angle tracking for the planar snake robot has been executed through SMC as shown in Fig. 2.1. The state vector for the constraint system (2.1a) is $\bar{\mathbf{x}} \in \mathbb{R}^8$ whereas the input vector is $\mathbf{u}_\sigma \in \mathbb{R}^2$. The sliding surfaces for the underactuated system have been chosen as

$$\boldsymbol{\sigma}(\bar{\mathbf{x}}) = \begin{bmatrix} \sigma_1(\bar{\mathbf{x}}) \\ \sigma_2(\bar{\mathbf{x}}) \end{bmatrix} = \begin{bmatrix} \dot{\tilde{\theta}}_n + K_n \tilde{\theta}_n \\ \dot{\lambda} + K_v \tilde{v}_t \end{bmatrix}, \tag{2.9}$$

where $K_n > 0$ and $K_v > 0$ are design parameters chosen such that the dynamics of the sliding manifold is stable. The reduced order dynamics on the sliding manifold $\boldsymbol{\sigma}(\bar{\mathbf{x}}) = 0$ yields

$$\dot{\tilde{\theta}}_n = -K_n \tilde{\theta}_n,$$
$$\tilde{v}_t = \frac{\dot{\lambda}}{K_v}.$$

For $K_n > 0$, the velocity error $\tilde{\theta}_n$ converges exponentially to the origin. On the other hand, for a sufficiently large $K_v > 0$, the velocity error $\tilde{v}_t$ can be made sufficiently small. To design the reaching law for the sliding surface (2.9), the system dynamics on the sliding manifold is obtained to be

$$\dot{\sigma}_1 = \ddot{\tilde{\theta}}_n + K_n \dot{\tilde{\theta}}_n = (\hat{\psi}_1 - \ddot{\theta}_{\text{ref}} + K_n \dot{\tilde{\theta}}_n + \psi_2 u_\lambda + \psi_3 u_{\phi_0} + \tilde{\psi}_{\theta_n}, \tag{2.10a}$$
$$\dot{\sigma}_2 = \ddot{\lambda} + K_v \dot{\tilde{v}}_t = K_v(f_2 - \dot{v}_{\text{ref}}) + u_\lambda + \tilde{\psi}_{v_t}. \tag{2.10b}$$

The above system can be represented in vector form as

$$\dot{\boldsymbol{\sigma}} = \mathbf{f}_\sigma + \mathbf{g}_\sigma \mathbf{u}_\sigma + \tilde{\mathbf{f}}_\sigma, \tag{2.11}$$

where

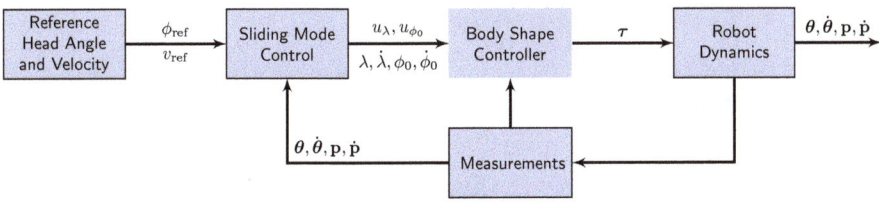

Fig. 2.1 Block diagram of control law for SMC

$$\mathbf{f}_\sigma = \begin{bmatrix} (\hat{\psi}_1 - \ddot{\theta}_{\text{ref}} + K_n \dot{\bar{\theta}}_n) \\ K_v(f_2 - \dot{v}_{\text{ref}}) \end{bmatrix}, \quad \mathbf{g}_\sigma = \begin{bmatrix} \psi_2 & \psi_3 \\ 1 & 0 \end{bmatrix},$$
$$\mathbf{u}_\sigma = \begin{bmatrix} u_\lambda & u_{\phi_0} \end{bmatrix}^T, \quad \tilde{\mathbf{f}}_\sigma = \begin{bmatrix} \tilde{\psi}_{\theta_n} & \tilde{\psi}_{v_t} \end{bmatrix}^T.$$
(2.12)

Assumption 2.1 The upper bound on the uncertainty in the system (2.11) is known a priori and is given as $||\tilde{\mathbf{f}}_\sigma|| \leqslant \rho(\boldsymbol{\theta}, \dot{\boldsymbol{\theta}}, \mathbf{p}, \dot{\mathbf{p}}, \lambda, \dot{\lambda}, \phi_0, \dot{\phi}_0)$, where $||.||$ represents vector 2-norm. □

2.3.1 Sliding-Mode Control Law

Once the reduced order system is on the sliding surface, it is desired that it stays on the surface. The control law for the system (2.11) is composed of two part. The first part, called *Equivalent Control*, stabilize the system on sliding surface to the origin in the absence of uncertainties [6]. The second part, called *Switching Control* ensures finite time reachability [9] to the sliding surface in the presence of uncertainties. Thus, the equivalent control input is computed by equating $\dot{\sigma}(\bar{\mathbf{x}}) = \mathbf{0}$ in the absence of uncertainty ($\tilde{\mathbf{f}}_\sigma = \mathbf{0}$) and can be expressed as

$$\mathbf{u}_{\sigma_{\text{eq}}} = -(\mathbf{g}_\sigma)^{-1} \mathbf{f}_\sigma. \quad (2.13)$$

To ensure the finite-time reachability of the system states to the sliding surface, a switching term is added in the control input [6] as

$$\mathbf{u}_{\sigma_{\text{sw}}} = -(\mathbf{g}_\sigma)^{-1} \eta \text{sign}(\boldsymbol{\sigma}). \quad (2.14)$$

where η is a positive constant design parameter called *switching gain*. The total control effort from SMC can be expressed by combining (2.13) and (2.14) as [3]

$$\mathbf{u}_\sigma = \mathbf{u}_{\sigma_{\text{eq}}} + \mathbf{u}_{\sigma_{\text{sw}}}. \quad (2.15)$$

The following subsection details the stability analysis of the constraint system acted upon by the proposed control law.

2.3.2 Stability Analysis

A condition that the control law (2.15) exists, is that $\mathbf{g}_\sigma$ is a non-singular matrix which can be assured as $\det(\mathbf{g}_\sigma) \neq 0$. The closed loop stability results are stated in the form of following theorem.

Theorem 2.1 *The input matrix $\mathbf{g}_\sigma$ of the constraint system (2.11) is invertible for the snake robot with parameters as given in Table 2.1.*

2.3 Sliding-Mode Control for Planar Snake Robot

Theorem 2.2 *Considering $\mathbf{g}_\sigma$ is non-singular, the constraint system (2.11) is finite-time stable using control law (2.15) if there exists a gain η such that $\eta > \rho$.*

Proof The first part of the proof provides a sufficient condition for the non-singularity of the input matrix $\mathbf{g}_\sigma$ as in Theorem 2.1. The stability of the system subject to the proposed control law as illustrated in Theorem 2.2 is presented in the second part of the proof.

2.3.2.1 Non-singularity of the Input Matrix $\mathbf{g}_\sigma$

The determinant of input matrix $\mathbf{g}_\sigma$ given in (2.12) can be obtained as

$$\det(\mathbf{g}_\sigma) = \psi_3.$$

This infers that the singularity of the input matrix $\mathbf{g}_\sigma$ depends on ψ_3. Therefore, considering the structure of ψ_3 (1.34a), the denominator of the same is given as

$$\mathbf{e}^T \mathbf{M} \mathbf{e} > 0, \tag{2.16}$$

as the mass matrix $\mathbf{M}$ of the snake robot is a positive definite matrix and $\mathbf{e}$ is a non-zero vector. Furthermore, the numerator can be written as

$$\mathbf{e}^T \mathbf{M} \mathbf{H} \mathbf{b}_1 = \mathbf{e}^T \mathbf{M} \bar{\mathbf{e}}, \tag{2.17}$$

where $\bar{\mathbf{e}}$ is given as

$$\bar{\mathbf{e}} = \begin{bmatrix} 1 & 1 & 1 & \ldots & 1 & 0 \end{bmatrix}^T \in \mathbb{R}^n.$$

Moreover, vectors $\mathbf{e}$ and $\bar{\mathbf{e}}$ can be expressed as

$$\mathbf{e} = \begin{bmatrix} \hat{\mathbf{e}}^T & 1 \end{bmatrix}^T, \quad \bar{\mathbf{e}} = \begin{bmatrix} \hat{\mathbf{e}}^T & 0 \end{bmatrix}^T, \tag{2.18}$$

where

$$\hat{\mathbf{e}} = \begin{bmatrix} 1 & 1 & 1 & \ldots & 1 \end{bmatrix}^T \in \mathbb{R}^{n-1}.$$

Further, the mass matrix $\mathbf{M}$ can be decomposed as

$$\mathbf{M} = \begin{bmatrix} \mathbf{M}_1 & \mathbf{M}_2 \\ \mathbf{M}_2^T & M_n \end{bmatrix}, \tag{2.19}$$

where $\mathbf{M}_1 > \mathbf{0} \in \mathbb{R}^{(n-1)\times(n-1)}$, $\mathbf{M}_2 \in \mathbb{R}^{(n-1)\times 1}$, $M_n \in \mathbb{R}^+$. Similarly, the matrix $\mathbf{V}$ can be partitioned as

$$\mathbf{V} = \begin{bmatrix} \mathbf{V}_1 & \mathbf{V}_2 \\ \mathbf{V}_2^T & V_n \end{bmatrix}, \quad (2.20)$$

where $\mathbf{V}_1 > \mathbf{0} \in \mathbb{R}^{(n-1)\times(n-1)}$, $\mathbf{V}_2 \in \mathbb{R}^{(n-1)\times 1}$, $V_n \in \mathbb{R}^+$. The numerator of ψ_3 (2.17) can be written as

$$\mathbf{e}^T \mathbf{M} \bar{\mathbf{e}} = \begin{bmatrix} \hat{\mathbf{e}}^T & 1 \end{bmatrix} \begin{bmatrix} \mathbf{M}_1 & \mathbf{M}_2 \\ \mathbf{M}_2^T & M_n \end{bmatrix} \begin{bmatrix} \hat{\mathbf{e}} \\ 0 \end{bmatrix},$$

$$\Rightarrow \mathbf{e}^T \mathbf{M} \bar{\mathbf{e}} = \hat{\mathbf{e}}^T \mathbf{M}_1 \hat{\mathbf{e}} + \mathbf{M}_2^T \hat{\mathbf{e}}, \quad (2.21)$$

For $\mathbf{M}_2^T \hat{\mathbf{e}} \geq 0$, $\mathbf{e}^T \mathbf{M} \bar{\mathbf{e}}$ can be assured to be positive for all θ as $\hat{\mathbf{e}}^T \mathbf{M}_1 \hat{\mathbf{e}} > 0$. Although, the same cannot be confirmed for $\mathbf{M}_2^T \hat{\mathbf{e}} < 0$. Hence, utilizing (2.18) and (2.19) in (2.16), one can deduce

$$\hat{\mathbf{e}}^T \mathbf{M}_1 \hat{\mathbf{e}} + 2\mathbf{M}_2^T \hat{\mathbf{e}} + M_n > 0,$$

$$\Rightarrow \mathbf{M}_2^T \hat{\mathbf{e}} > -\frac{(\hat{\mathbf{e}}^T \mathbf{M}_1 \hat{\mathbf{e}} + M_n)}{2}. \quad (2.22)$$

Utilizing the lower bound of $\mathbf{M}_2^T \hat{\mathbf{e}}$ (2.22) in (2.21) yields

$$\mathbf{e}^T \mathbf{M} \bar{\mathbf{e}} > \hat{\mathbf{e}}^T \mathbf{M}_1 \hat{\mathbf{e}} - \frac{(\hat{\mathbf{e}}^T \mathbf{M}_1 \hat{\mathbf{e}} + M_n)}{2},$$

$$\Rightarrow \mathbf{e}^T \mathbf{M} \bar{\mathbf{e}} > \frac{(\hat{\mathbf{e}}^T \mathbf{M}_1 \hat{\mathbf{e}} - M_n)}{2}. \quad (2.23)$$

Hence, the sufficient condition for the non-singularity of ψ_3 can be obtained from (2.23)

$$\hat{\mathbf{e}}^T \mathbf{M}_1 \hat{\mathbf{e}} - M_n > 0,$$

$$\Rightarrow \hat{\mathbf{e}}^T \mathbf{M}_1 \hat{\mathbf{e}} > M_n. \quad (2.24)$$

The matrix $\mathbf{M}_1$ can be expressed as

$$\mathbf{M}_1 = J\mathbf{I}_{(n-1)} + ml^2 \bar{\mathbf{S}}_\theta \mathbf{V}_1 \bar{\mathbf{S}}_\theta + ml^2 \bar{\mathbf{C}}_\theta \mathbf{V}_1 \bar{\mathbf{C}}_\theta, \quad (2.25)$$

where

$$\bar{\mathbf{S}}_\theta = \text{diag}(\mathbf{sin}\bar{\theta}) \in \mathbb{R}^{(n-1)\times(n-1)},$$
$$\mathbf{sin}\bar{\theta} = \begin{bmatrix} \sin(\theta_1) & \sin(\theta_2) & \ldots & \sin(\theta_{n-1}) \end{bmatrix}^T \in \mathbb{R}^{(n-1)},$$
$$\bar{\mathbf{C}}_\theta = \text{diag}(\mathbf{cos}\bar{\theta}) \in \mathbb{R}^{(n-1)\times(n-1)},$$
$$\mathbf{cos}\bar{\theta} = \begin{bmatrix} \cos(\theta_1) & \cos(\theta_2) & \ldots & \cos(\theta_{n-1}) \end{bmatrix}^T \in \mathbb{R}^{(n-1)}.$$

2.3 Sliding-Mode Control for Planar Snake Robot

Hence, the lower bound of $\hat{\mathbf{e}}^T \mathbf{M}_1 \hat{\mathbf{e}}$ can be computed as

$$\hat{\mathbf{e}}^T \mathbf{M}_1 \hat{\mathbf{e}} = J\hat{\mathbf{e}}^T \mathbf{I}_{(n-1)}\hat{\mathbf{e}} + ml^2 \big(\hat{\mathbf{e}}^T \bar{\mathbf{S}}_\theta \mathbf{V}_1 \bar{\mathbf{S}}_\theta \hat{\mathbf{e}} + \hat{\mathbf{e}}^T \bar{\mathbf{C}}_\theta \mathbf{V}_1 \bar{\mathbf{C}}_\theta \hat{\mathbf{e}}\big),$$
$$\Rightarrow \hat{\mathbf{e}}^T \mathbf{M}_1 \hat{\mathbf{e}} = J(n-1) + ml^2 \big(\sin\bar{\theta}^T \mathbf{V}_1 \sin\bar{\theta} + \cos\bar{\theta}^T \mathbf{V}_1 \cos\bar{\theta},\big)$$
$$\Rightarrow \hat{\mathbf{e}}^T \mathbf{M}_1 \hat{\mathbf{e}} \geq J(n-1) + ml^2 \lambda_{\min}(\mathbf{V}_1)\big(\|\sin\bar{\theta}\|^2 + \|\cos\bar{\theta}\|^2\big),$$
$$\Rightarrow \hat{\mathbf{e}}^T \mathbf{M}_1 \hat{\mathbf{e}} \geq (n-1)\big(J + ml^2 \lambda_{\min}(\mathbf{V}_1)\big). \quad (2.26)$$

Moreover, the value of M_n can be expressed as

$$M_n = J + ml^2 V_n. \quad (2.27)$$

Thereupon, (2.26) and (2.27) can be utilized in condition (2.24) as

$$(n-1)\big(J + ml^2 \lambda_{\min}(\mathbf{V}_1)\big) > J + ml^2 V_n,$$
$$(n-2)J > ml^2 \big(V_n - (n-1)\lambda_{\min}(\mathbf{V}_1)\big). \quad (2.28)$$

Finally, for m, l and J given in Table 2.1, the value of n satisfying (2.28) is found to be

$$0.0016(n-2) > 0.0041,$$
$$n > 4.5625.$$

From analysis of the constant $\mathbf{V}$ matrix, one can deduce

$$\left.\begin{array}{l} 0.5 \geq \lambda_{min}(\mathbf{V}_1) \to 0 \\ 0.5 \leq V_n \to 1 \end{array}\right\} \quad n = 2, 3, \ldots, \infty. \quad (2.29)$$

Furthermore, the moment of inertia J is related to the mass of the link m and link length $2l$ based upon the link profile and the distribution of mass over the link. Consequently, one can design the link for a snake robot with n number of links satisfying (2.28) as

$$n > \frac{ml^2}{J} + 2. \quad (2.30)$$

Therefore, (2.30) presents a sufficient condition toward the choice of robot parameters so that Theorem 2.1 is fulfilled [10].

This confirms that ψ_3 is non-zero for the physical parameters given in Table 2.1 (where $n = 10$) thus making $\mathbf{g}_\sigma$ a non-singular matrix.

2.3.2.2 Lyapunov Stability

Let a Lyapunov function candidate of the form

$$V = \frac{1}{2}\boldsymbol{\sigma}^T\boldsymbol{\sigma}. \tag{2.31}$$

The derivative of the Lyapunov function candidate w.r.t time can be written as

$$\begin{aligned}\dot{V} &= \boldsymbol{\sigma}^T\dot{\boldsymbol{\sigma}} = \boldsymbol{\sigma}^T(\mathbf{f}_\sigma + \mathbf{g}_\sigma\mathbf{u}_\sigma + \tilde{\mathbf{f}}_\sigma), \\ &= \boldsymbol{\sigma}^T\big(\mathbf{f}_\sigma + \mathbf{g}_\sigma(\mathbf{u}_{\sigma_{eq}} + \mathbf{u}_{\sigma_{sw}}) + \tilde{\mathbf{f}}_\sigma\big), \\ &= \boldsymbol{\sigma}^T\big(-\eta\mathrm{sign}(\boldsymbol{\sigma}) + \tilde{\mathbf{f}}_\sigma\big). \end{aligned} \tag{2.32}$$

Utilizing Assumption 2.1 in the above derivation, one can write

$$\begin{aligned}\dot{V} &\leq -\eta\,\|\boldsymbol{\sigma}^T\| + \rho\,\|\boldsymbol{\sigma}^T\|, \\ &\leq -(\eta-\rho)\,\|\boldsymbol{\sigma}^T\|, \\ &= -(\eta-\rho)(V)^{1/2}. \end{aligned} \tag{2.33}$$

This proves that $\boldsymbol{\sigma}$ converges to $\mathbf{0}$ in finite time i.e. the finite time stability of the system (2.11) on application of control law (2.15) is achieved [3]. □

2.4 Adaptive Sliding-Mode Control for Planar Snake Robots

The control approach presented in the previous section presents a successful attempt to deal with uncertainties. Once such an attempt has been made, it is time to improve the SMC approach to extract improved performance. In-depth analysis of the SMC-based control law reveal, that the switching control law is the primary contributor to the overall control input to the robot. Being a mobile robot, it is always desirable to have the least control input, as that results in extended work-area and work-time. Hence, the following subsection presents an adaptation law, and the corresponding control approach has been schematically described in Fig. 2.2.

2.4.1 Adaptive Sliding-Mode Control Law

This section proposes an *adaptation law* to update the switching gain in a way, that the system error can be minimized keeping the switching gain as small as possible. Also, using an adaptation law, the control algorithm is thus freed of the requirement of uncertainty upper bound knowledge. Although the uncertainties are still required to be bounded.

2.4 Adaptive Sliding-Mode Control for Planar Snake Robots

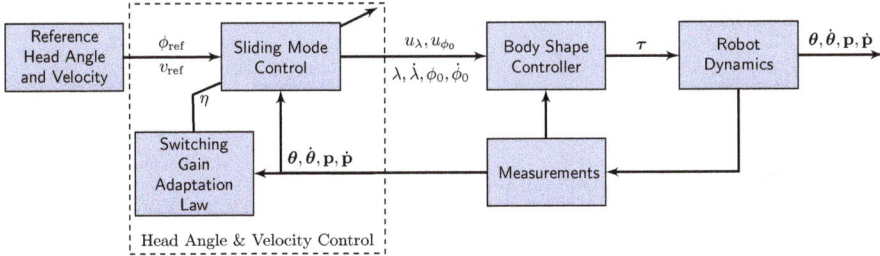

Fig. 2.2 Block diagram of control law for ASMC

2.4.1.1 Adaptation Law

As we have already highlighted, a constant value of η though ensures stability, results in a considerably high average control input as presented in Sect. 2.3. In the presence of time-varying uncertainty, this work proposes an adaptation law to update the value of the switching gain η that confirms stability of the constrained system. This approach considerably reduce the average control input to the robot. The adaptation law chosen is given as

$$\dot{\eta} = \begin{cases} \bar{\eta} \|\boldsymbol{\sigma}\| \text{sign}(\|\boldsymbol{\sigma}\| - \epsilon) & for \quad \eta > \mu \\ \mu & for \quad \eta \leq \mu \end{cases}, \quad (2.34)$$

where $\eta(0)$, $\bar{\eta}$, μ and ϵ are positive controller parameters to be chosen. The value μ has been set to restrict the gain from going to non-positive value. A stronger condition to be chosen henceforth is $\eta(t) > \mu$ for all $t > 0$. The parameter ϵ defines a region around the sliding surface that indicates closeness to the surface which will stop further variation in the switching gain as establishment of *ideal sliding mode* is not possible in real application.

Lemma 2.1 *For a system defined in* (1.8) *with sliding variable* ($\boldsymbol{\sigma}$) *dynamics* (2.11) *controlled by the* (1.31), (2.15) *and* (2.34), *the switching gain η has an upper bound i.e. there exists a positive constant η^* such that [7]*

$$\eta(t) \leq \eta^*, \forall t > 0.$$

Lemma 2.1 has been utilized to ascertain the *finite-time convergence* of the system states to the sliding manifold [4].

2.4.2 Stability Analysis

Theorem 2.3 *The sliding system (2.11) controlled by (2.15) and (2.34), there exists a finite time $t_F > 0$ so that a real sliding mode is established for all $t \geq t_F$.*

Proof Let a Lyapunov function candidate of the form

$$V = \frac{1}{2}\sigma^T\sigma + \frac{1}{2\gamma}(\eta - \eta^*)^2. \tag{2.35}$$

The derivative of the Lyapunov function candidate w.r.t time can be expressed as

$$\begin{aligned}\dot{V} &= \sigma^T\dot{\sigma} + \frac{1}{\gamma}(\eta - \eta^*)\dot{\eta}, \\ &= \sigma^T(\mathbf{f}_\sigma + \mathbf{g}_\sigma\mathbf{u}_\sigma + \tilde{\mathbf{f}}_\sigma) + \frac{1}{\gamma}(\eta - \eta^*)\dot{\eta}, \\ &= \sigma^T\big(-\eta\operatorname{sign}(\sigma) + \tilde{\mathbf{f}}_\sigma\big) + \frac{1}{\gamma}(\eta - \eta^*)\dot{\eta}, \\ &\leq (\hat{f}_\sigma - \eta)\,\|\sigma\| + \frac{1}{\gamma}(\eta - \eta^*)\bar{\eta}\,\|\sigma\|\,\operatorname{sign}(\|\sigma\| - \epsilon).\end{aligned} \tag{2.36}$$
$$\tag{2.37}$$

Adding and subtracting a term $\eta^*\,\|\sigma\|$ in (2.36), we get

$$\dot{V} = (\hat{f}_\sigma - \eta^*)\,\|\sigma\| + (\eta - \eta^*)\Big(-\|\sigma\| + \frac{\bar{\eta}}{\gamma}\,\|\sigma\|\,\operatorname{sign}(\|\sigma\| - \epsilon)\Big). \tag{2.38}$$

Introducing a positive value $\beta_k > 0$ in (2.38) as

$$\dot{V} = (\hat{f}_\sigma - \eta^*)\,\|\sigma\| - \beta_k\,\|\eta - \eta^*\| + (\eta - \eta^*)\Big(-\|\sigma\| \\ + \frac{\bar{\eta}}{\gamma}\,\|\sigma\|\,\operatorname{sign}(\|\sigma\| - \epsilon)\Big) + \beta_k\,\|\eta - \eta^*\|. \tag{2.39}$$

From Lemma 2.1, there always exist a η^* such that $\eta - \eta^* < 0\;\forall\,t > 0$. This results in to

$$\begin{aligned}\dot{V} &= -(-\hat{f}_\sigma + \eta^*)\,\|\sigma\| - \beta_k\,\|\eta - \eta^*\| - \Big(-\|\sigma\| + \frac{\bar{\eta}}{\gamma}\,\|\sigma\|\,\operatorname{sign}(\|\sigma\| - \epsilon) - \beta_k\Big)\,\|\eta - \eta^*\|, \\ &= -\beta_\sigma\,\|\sigma\| - \beta_k\,\|\eta - \eta^*\| - \zeta,\end{aligned} \tag{2.40}$$

where $\beta_\sigma = (-\hat{f}_\sigma + \eta^*) > 0$ and $\zeta = \big(-\|\sigma\| + \frac{\bar{\eta}}{\gamma}\,\|\sigma\|\,\operatorname{sign}(\|\sigma\| - \epsilon) - \beta_k\big)\,\|\eta - \eta^*\|$.

2.4 Adaptive Sliding-Mode Control for Planar Snake Robots

Continuing

$$\dot{V} = -\beta_\sigma \sqrt{2} \frac{\|\sigma\|}{\sqrt{2}} - \beta_k \sqrt{2\gamma} \frac{\|\eta - \eta^*\|}{\sqrt{2\gamma}} - \zeta,$$

$$\leqslant -\min\{\beta_\sigma \sqrt{2}, \beta_k \sqrt{2\gamma}\} \left(\frac{\|\sigma\|}{\sqrt{2}} + \frac{\|\eta - \eta^*\|}{\sqrt{2\gamma}} \right) - \zeta. \tag{2.41}$$

The Lyapunov function V in (2.35) can be written as

$$V = \frac{1}{2} \sigma^T \sigma + \frac{1}{2\gamma}(\eta - \eta^*)^2 = \left\{ \frac{\|\sigma\|}{\sqrt{2}} \right\}^2 + \left\{ \frac{\|\eta - \eta^*\|}{\sqrt{2\gamma}} \right\}^2. \tag{2.42}$$

Utilizing *triangle inequality*, one can write

$$\left\{ \frac{\|\sigma\|}{\sqrt{2}} + \frac{\|\eta - \eta^*\|}{\sqrt{2\gamma}} \right\}^2 \geq \left\{ \frac{\|\sigma\|}{\sqrt{2}} \right\}^2 + \left\{ \frac{\|\eta - \eta^*\|}{\sqrt{2\gamma}} \right\}^2,$$

$$\frac{\|\sigma\|}{\sqrt{2}} + \frac{\|\eta - \eta^*\|}{\sqrt{2\gamma}} \geq \left[\left\{ \frac{\|\sigma\|}{\sqrt{2}} \right\}^2 + \left\{ \frac{\|\eta - \eta^*\|}{\sqrt{2\gamma}} \right\}^2 \right]^{1/2}. \tag{2.43}$$

Multiplying both sides with -1, the inequality can be expressed as

$$-\left\{ \frac{\|\sigma\|}{\sqrt{2}} + \frac{\|\eta - \eta^*\|}{\sqrt{2\gamma}} \right\} \leq -\left[\left\{ \frac{\|\sigma\|}{\sqrt{2}} \right\}^2 + \left\{ \frac{\|\eta - \eta^*\|}{\sqrt{2\gamma}} \right\}^2 \right]^{1/2},$$

$$-\left\{ \frac{\|\sigma\|}{\sqrt{2}} + \frac{\|\eta - \eta^*\|}{\sqrt{2\gamma}} \right\} \leq -V^{1/2}. \tag{2.44}$$

Using (2.44) in (2.41), one can obtain

$$\dot{V} \leqslant -\beta V^{1/2} - \zeta. \tag{2.45}$$

Case 1: if $\|\sigma\| > \epsilon$, ζ is positive if,

$$-\|\sigma\| + \frac{\bar{\eta}}{\gamma} - \beta_k > 0 \Rightarrow \gamma < \frac{\bar{\eta}\epsilon}{\epsilon + \beta_k}. \tag{2.46}$$

With the condition (2.42) holding, the time derivative of the Lyapunov function becomes

$$\dot{V} \leqslant -\beta V^{1/2}. \tag{2.47}$$

Case 2: if $\|\sigma\| \leqslant \epsilon$, ζ can be negative and it is not possible to prove negative-definiteness of $\dot{V}$.

This proves that $\|\boldsymbol{\sigma}\|$ converges to a region ϵ in finite time. But once inside ϵ, the trajectory of the sliding variable can not be ascertained. If at some t_{F_1} the sliding state goes beyond ϵ, i.e. $\|\boldsymbol{\sigma}(t_{F_1})\| > \epsilon$, there exists another finite time t_{F_2} when the states will re-converge to the region ϵ. This proves the establishment of a real sliding mode [4]. □

2.5 Simulation Results

In this section, the simulation scenario and results for the SMC and ASMC strategies on the planar snake robot have been presented. The numerical simulations have been performed in MATLAB/Simulink environment.

2.5.1 Simulation Scenario

The parametric values considered in the simulation are inspired by a physically realized robot [11] which also allows a performance comparison of the proposed methodology with respect to the state-of-the-art approach. The physical specifications of the snake robot considered are given in Table 2.1. Uncertainties have been considered in the tangential and normal friction coefficients between the link surface and the plane below. The unknown variation in the friction coefficients c_t and c_n have been chosen as such, to imitate a change in the ground condition as a function of time. The time-varying friction coefficients shown in Fig. 2.3 are generated as

$$c_t = \begin{cases} -8 \times 10^{-7} t^3 + 1.2 \times 10^{-4} t^2 + 0.1, t \in [0\ 100] \\ 0.5, t \in [100\ 200] \\ 8 \times 10^{-7} t^3 - 6 \times 10^{-4} t^2 + 0.144 t - 10.7, t \in [200\ 300] \end{cases},$$

$$c_n = \begin{cases} -4.8 \times 10^{-6} t^3 + 7.2 \times 10^{-4} t^2 + 0.6, t \in [0\ 100] \\ 3, t \in [100\ 200] \\ 4.8 \times 10^{-6} t^3 - 0.0036 t^2 + 0.864 t - 64.2, t \in [200\ 300] \end{cases}.$$

Table 2.1 System parameters

Parameter	Numerical value
n	10
m	1 kg
$2l$	0.14 m
J	1.6×10^{-3} kgm^2

2.5 Simulation Results

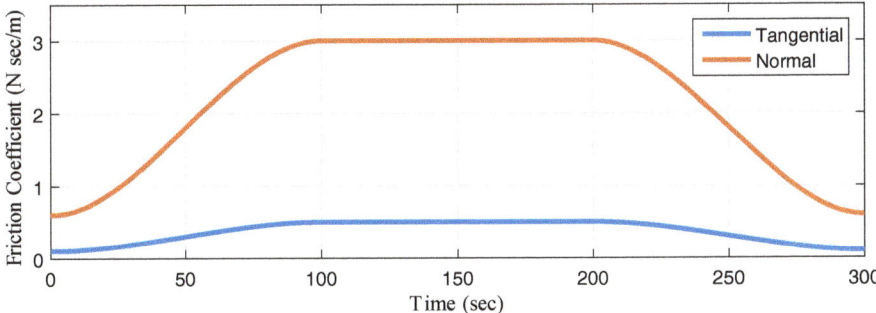

Fig. 2.3 Time-varying friction coefficient

Table 2.2 Reference and nominal values for SMC

Parameter	Numerical value
θ_{ref}	$-\pi/4$ rad
v_{ref}	0.05 m/s^2
$\hat{c}_t$	1
$\hat{c}_n$	4
α	$30\pi/180$ rad
δ	$72\pi/180$ rad

The reference velocity and head-angle along with the nominal friction coefficients and the constant gain parameters are given in Table 2.2. The simulation was done in MATLAB/Simulink environment and was run for 300 s. All the simulations have been performed with *non-zero* initial conditions in velocity and head-angle errors.

2.5.2 Results for Sliding-Mode Control

The various control gains and design parameters of the proposed SMC and the existing controller are shown in Table 2.3 [3].

The control parameter η has been chosen to be greater than the upper bound on the uncertainties, i.e. $\eta > \rho$. This is a strict condition that ensures the stability of the system dynamics on the VHC manifold.

Identical conditions were used to analyze the performance of the proposed control law and the control law presented in [11]. The trajectory traversed by the robot in global space employing the two control approaches is shown in Fig. 2.4. The robot utilizing SMC-based approach can be seen to be traversing a longer distance. In comparing the body-shape controller as shown in Fig. 2.5, we see that the proposed controller has an oscillatory response as a result of the switching control. The sliding surfaces designed for the SMC-based control approach are given in Fig. 2.6 which show finite-time reachability of the sliding function. The tangential velocity error and

Table 2.3 Control parameters for SMC

Parameter	Numerical value
$\mathbf{K}_P$	$10I_{n-1}$
$\mathbf{K}_D$	$10I_{n-1}$
K_n	1
K_v	12
η	12
ϵ	0.1
K_N	10
k_1	1
k_2	1
K_z	30
K_λ	12

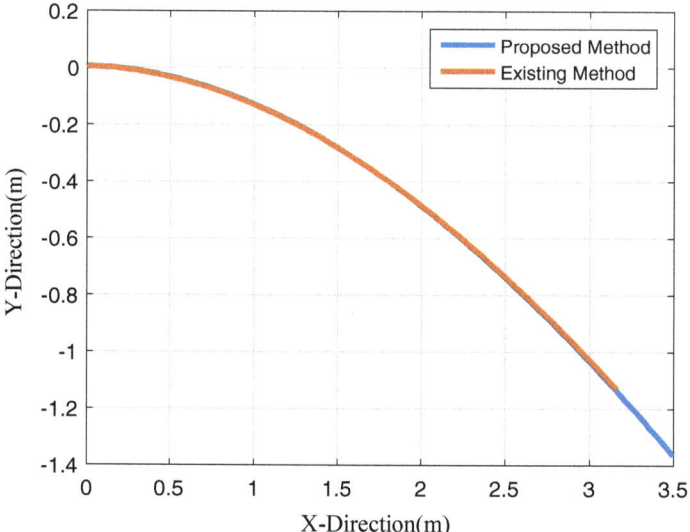

Fig. 2.4 Global trajectory

corresponding frequency for both the control laws are shown in Fig. 2.7 and Fig. 2.8, respectively. These responses affirm the superior velocity tracking capability of the proposed controller in presence of bounded uncertainties. Similarly, the head-angle error and corresponding angular offset value for the two control approaches are shown in Fig. 2.9 and Fig. 2.10 respectively. In this case as well, the SMC-based control method marginally outperforms the existing controller. Finally, Fig. 2.11 plots the norm of all the control inputs required by the SMC-based approach and the existing control method [11]. The initial high control input for the existing method can be associated with the singular perturbation-based head angle control law. The higher

2.5 Simulation Results

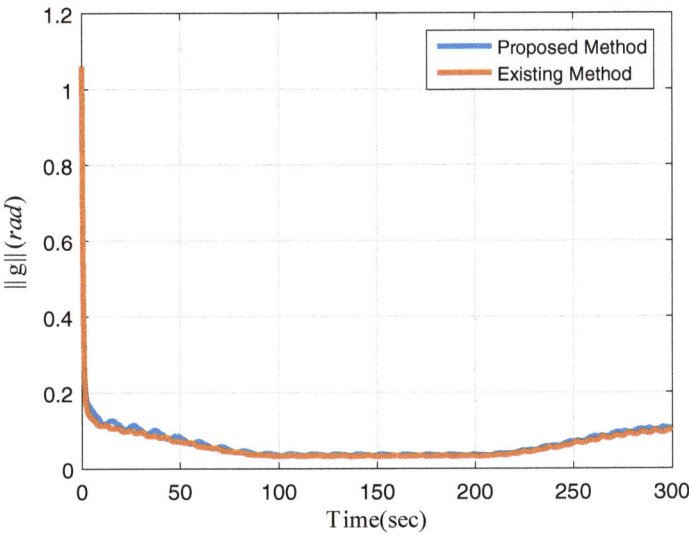

Fig. 2.5 Norm of VHCs

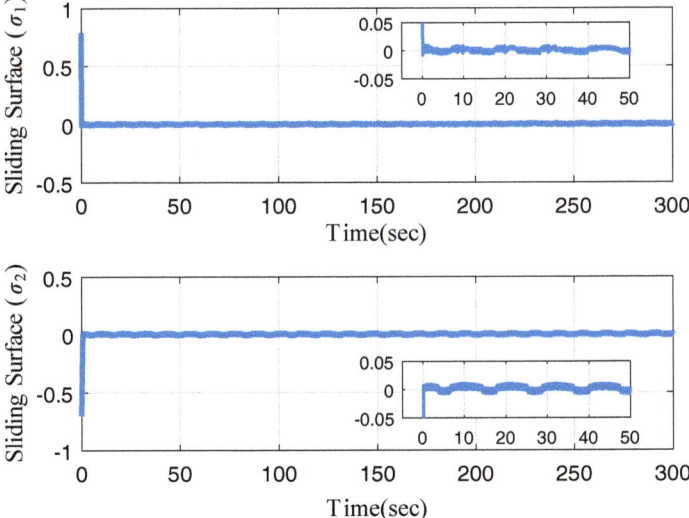

Fig. 2.6 Sliding surfaces

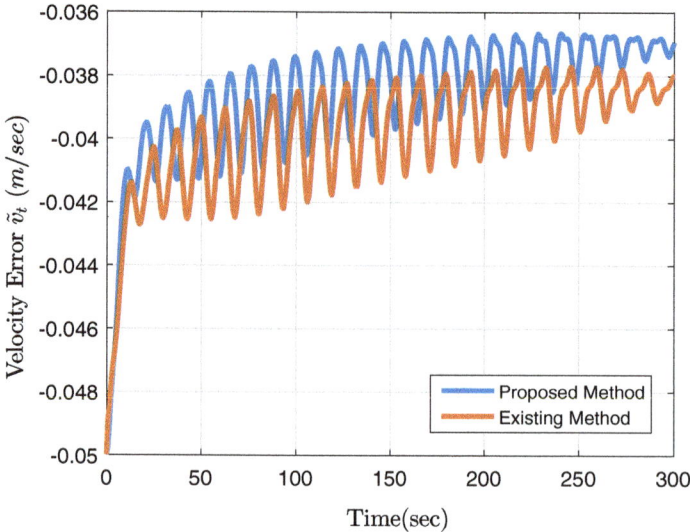

Fig. 2.7 Tangential velocity error

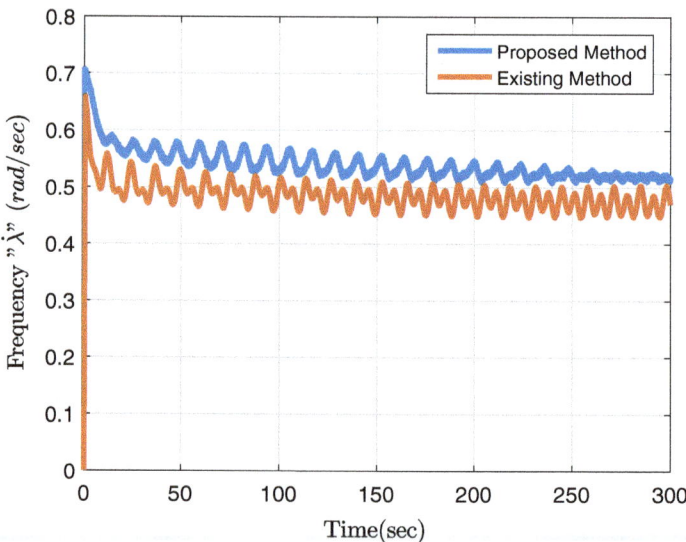

Fig. 2.8 Gait function frequency

2.5 Simulation Results

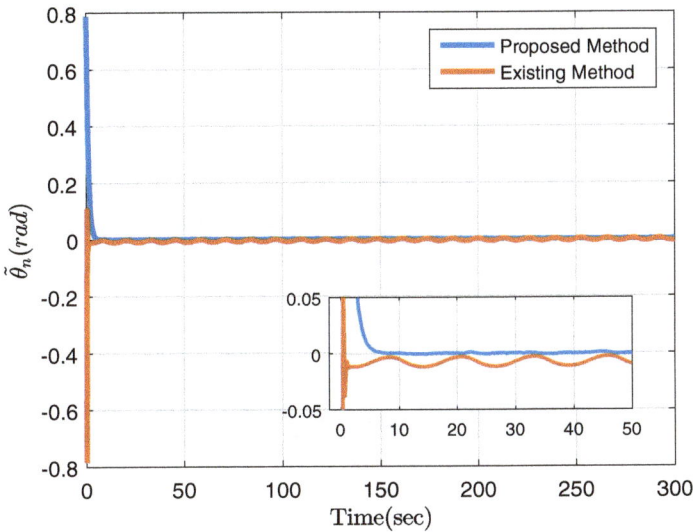

Fig. 2.9 Head-angle error

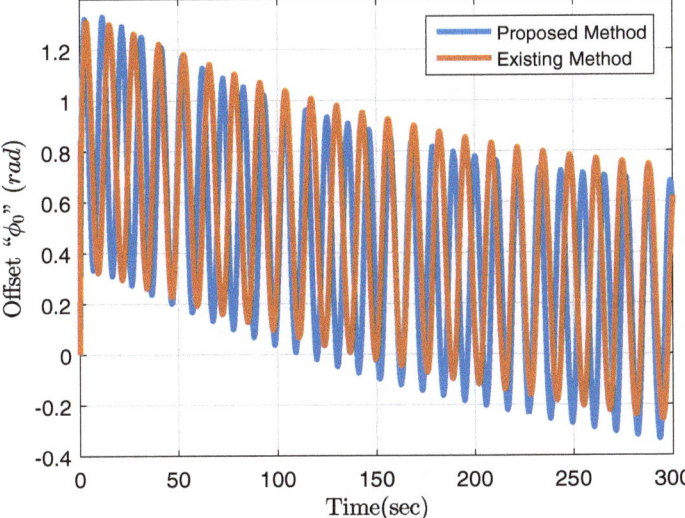

Fig. 2.10 Gait function offset

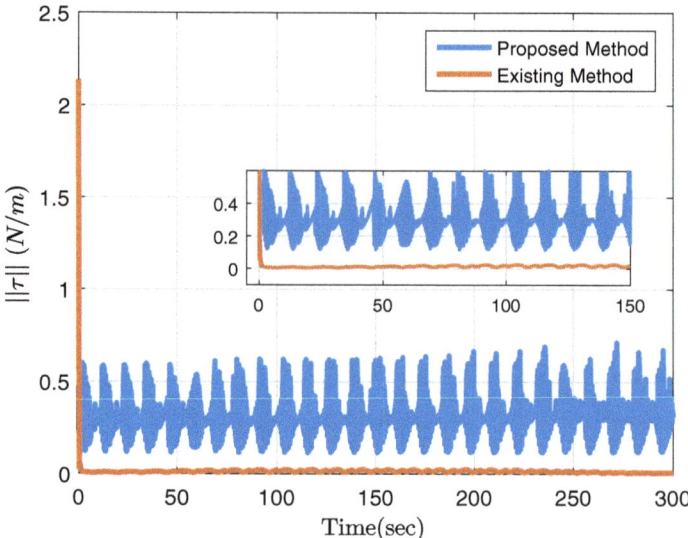

Fig. 2.11 Norm of control effort

Table 2.4 Adaptation law parameters for ASMC

Parameter	Numerical value
μ	1.5
$\bar{\eta}$	0.5
T_e	0.001 s

average torque requirement for the proposed controller is due to the switching control law. But, we observe that neither the frequency nor the torque values for the proposed method have reached anywhere near the saturating values. The peaks in Figs. 2.5, 2.6, 2.7, 2.8, 2.9 can be associated with the initial non-zero error and the switching control law.

2.5.3 Results for Adaptive Sliding-Mode Control

The simulation environment has been kept identical to the SMC-based approach for the sake of a fair comparison. The robot specifications are given in Table 2.1, whereas trajectory references and nominal parameters are cited in Table 2.2. All the control gains other than η can be referred to from Table 2.3.

The control parameters regarding the adaptation law of the switching gain η are presented in Table 2.4 [4].

The sliding region ϵ has been chosen as $\epsilon = 4\,\eta(t)T_e$ in simulation. The choice of the control variables ϵ is crucial as the stability of the system highly depends on

2.5 Simulation Results

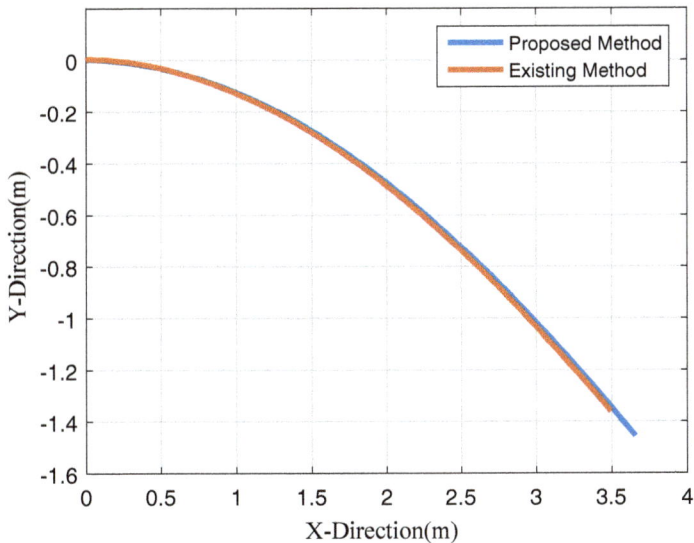

Fig. 2.12 Global trajectory of the snake robot for ASMC

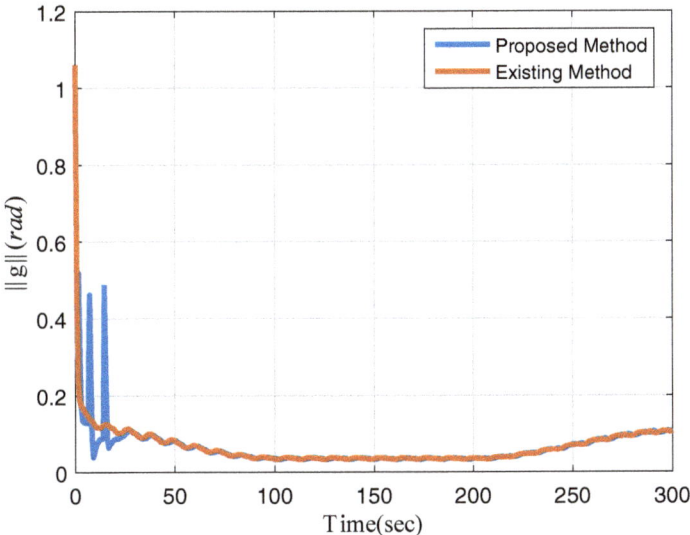

Fig. 2.13 Norm of VHCs

it. If ϵ is chosen to be a very small value, the system might never converge to the band and may result in an infinite gain. On the other hand, a large value of ϵ might reduce the accuracy of the control law.

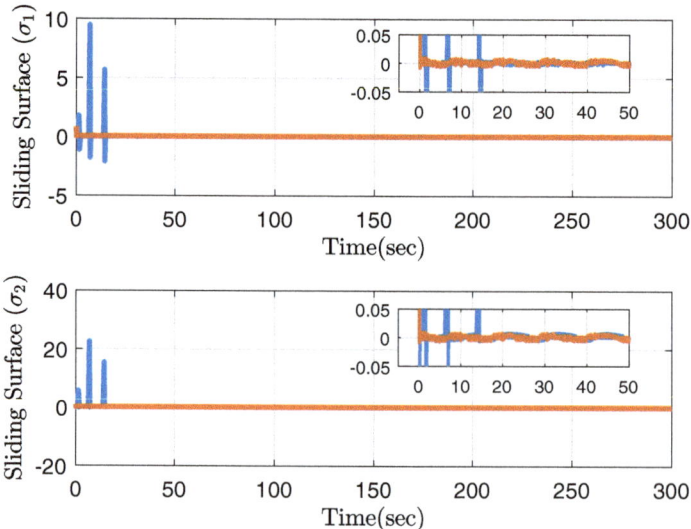

Fig. 2.14 Sliding surfaces for ASMC

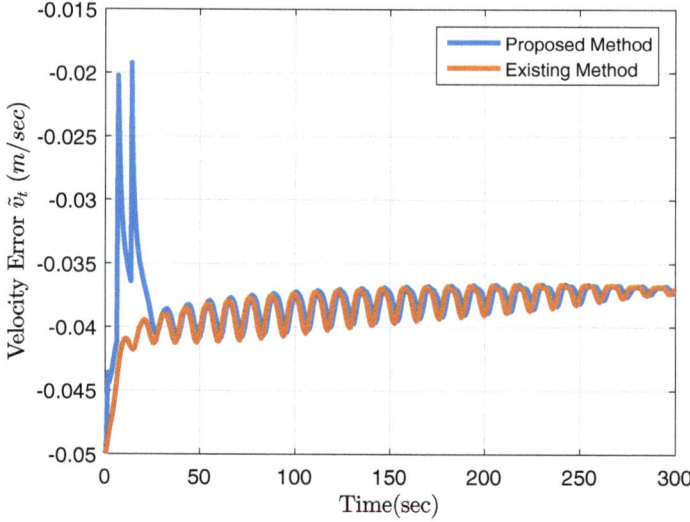

Fig. 2.15 Tangential velocity error for ASMC

2.5 Simulation Results

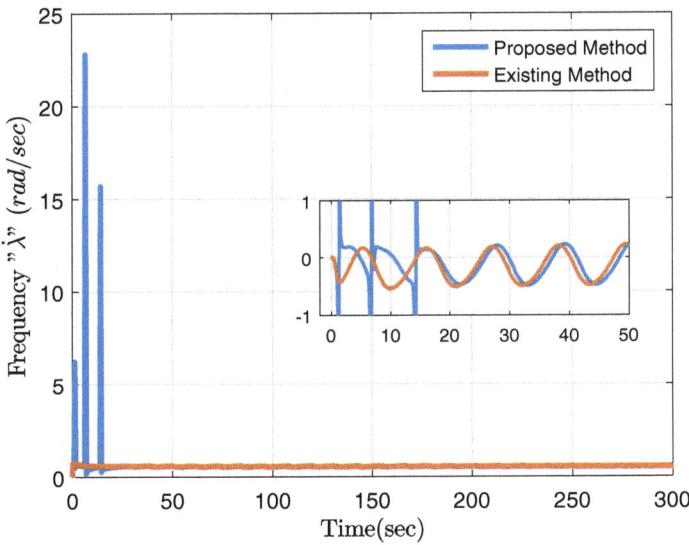

Fig. 2.16 Frequency of the gait function for ASMC

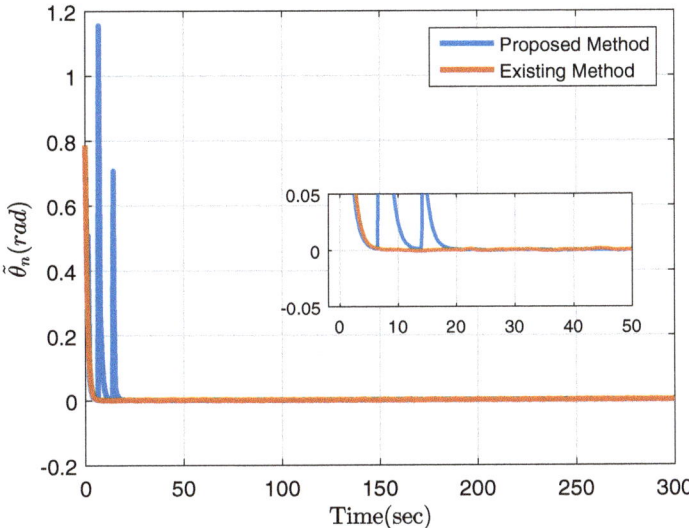

Fig. 2.17 Global head-angle error for ASMC

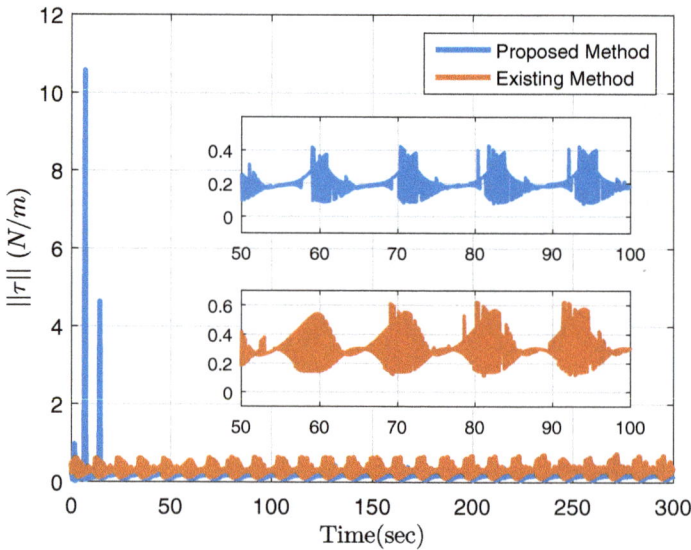

Fig. 2.18 Norm of the control effort for ASMC

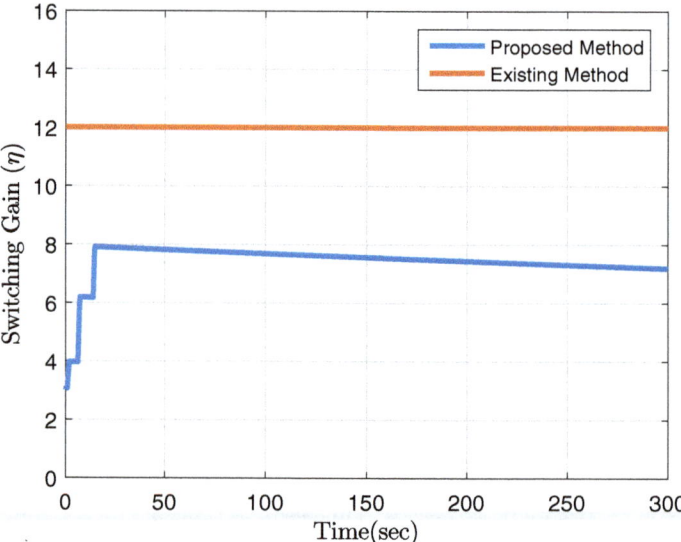

Fig. 2.19 Switching gain for ASMC

2.5 Simulation Results

The global trajectory of the robot utilizing the proposed method is seen to be traversing a larger distance with improved heading than the existing method in Fig. 2.12 which ascertains better velocity tracking. The norm of the VHCs shown in Fig. 2.13 exhibit initial oscillations for the ASMC based approach owing to the adaptation of the switching gain. In the case of the sliding surfaces, the proposed approach achieves similar results to the existing method only with some initial peaks as shown in Fig. 2.14. The velocity error also shows initial peaks for the proposed scheme in Fig. 2.15. This figure also exhibits improved steady-state velocity tracking of the proposed method in comparison to the exiting approach. The corresponding gait frequency also exhibits initial peaks in Fig. 2.16. The surges though being high would go unnoticed as the actuators take time to adjust the same. The head-angle error as well shows similar transient peaks for the proposed scheme with improved steady-state response as shown in Fig. 2.17. The norm of the control input $\tau(t)$ shown in Fig. 2.18 exhibit initial peaks for the Adaptive-SMC based method but lower average value relative to the SMC-based approach. The amplitude of the control torque has gone down by approximately 33%. Lastly, the adaptive variation of the switching gain for the proposed approach and the constant gain for the existing approach have been presented in Fig. 2.19. Here, the adaptive law is responsible for choosing the switching gain large enough for finite-time stability. The constant switching gain η for the SMC (Sect. 2.3) has been chosen appropriately to achieve efficient tracking performance (Fig. 2.14).

2.5.4 Discussion

The velocity of the snake robot is controlled by the frequency $\dot{\lambda}$ of the serpenoid gait function. The actuation at the joints, acting as wave generators, lead to the undulation of the robot body. The maximum frequency and torque for the motors in the present work have been limited to be 2 rad/s and 7 N–m as in [11]. High values of torque amplitude and frequency can saturate the motor leading to hard nonlinearities and may even cause damage to the motors.

The control gains for the existing control method have been chosen as given in Table 2.3. It can be noted that gains have been kept at similar values for achieving a better comparison of the two control methods. The control law proposed in [11] doesn't consider the presence of bounded uncertainties during controller design, whereas the SMC-based proposed method does. Input–output linearization is an effective control method when the system is exactly known. Under uncertainties, exact cancellation is not achieved and control law may fail to provide satisfactory performance. Also, singular perturbation-based control law is known to demand high control input. The initial time of Fig. 2.11 shows a high value of control input for the existing method which can be associated with fast dynamics of the singular perturbation-based control law. The average control input is higher for the proposed method primarily due to the switching control part. Both head angle control and velocity control can be seen to be more effective through the proposed method than the control law proposed in [11] under similar gain conditions. The better velocity tracking can be associated with reaching a higher value of gait angle frequency which is evident from Figs. 2.7 and 2.8.

The switching gain has been found to be the primary contributor to the control effort applied by the motors to achieve the desired trajectory. The switching gain in the SMC-based method has to be chosen sufficiently larger than the uncertainty bound. On the other hand, the adaptive law results in a lower switching gain depending on time-varying uncertainty. The adaptive variation of gain results in improved velocity and head-angle tracking utilizing lower control input. All the performance plots show initial peaks which can be associated with the gain adaptation. These peaks do not posses any serious threat to the system and are within limits. The proposed control scheme results in initial peaks in all the parameters which can be associated with the initial adaptation phase. Despite the initial oscillations, the control input is throughout within the maximum limit that the actuators can deliver.

2.6 Summary

An SMC approach has been demonstrated in this chapter for effective head-angle and velocity tracking of a snake robot in the presence of system uncertainty. The body shape of the snake robot has been stabilized by regulating VHCs to zero using a feedback control law. The head-angle and velocity control are achieved on the reduced dynamics of the robot on the constraint manifold. Appropriate choice of sliding surfaces has been presented along with the corresponding reaching law and the switching control. The composite controller is tested by simulation with bounded uncertainty in ground friction. Time-varying friction coefficients have been applied to analyze the robustness of the proposed approach thus imitating the variation in ground conditions. The simulation shows effective tracking which is robust in the presence of the aforementioned uncertainties.

Consecutively, an adaptive-sliding-mode-based control law has been presented over the SMC-based approach by inducing an adaptation law for the switching gain. Incorporation of the adaptation law circumvents the requirement of any knowledge regarding the uncertainty bounds. Moreover, the overestimation of the switching gain in SMC has been solved through the adaptation law thus utilizing lower control input achieving improved tracking performance. This is supposed to utilize lesser battery power thus increasing the time duration and space over which the robot can work without recharging. Similar time-varying friction coefficients have been applied to analyze the robustness of the proposed approach which was found to be satisfactory.

As mentioned earlier, these control approaches require the uncertainties to be bounded and even the upper bound to be known especially for the SMC methodology. The applicability of the proposed schemes gets limited in situations where the aforementioned assumptions might not hold true. Hence, there is a scope for a different robust control approach that might address these issues and provide a more non-conservative approach as presented in the next chapter.

References

1. Slotine, J.J.E., Li, W., et al.: Applied Nonlinear Control, vol. 199. Prentice Hall, Englewood Cliffs, NJ (1991)
2. Khalil, H.K., Grizzle, J.: Nonlinear Systems, vol. 3. Prentice Hall, Upper Saddle River, NJ (2002)
3. Mukherjee, J., Mukherjee, S., Kar, I.N.: Sliding mode control of planar snake robot with uncertainty using virtual holonomic constraints. IEEE Robot. Autom. Lett. **2**(2), 1077–1084 (2017). https://doi.org/10.1109/LRA.2017.2657892
4. Mukherjee, J., Kar, I.N., Mukherjee, S.: Adaptive sliding mode control for head-angle and velocity tracking of planar snake robot. In: 2017 11th Asian Control Conference (ASCC), pp. 537–542 (2017). https://doi.org/10.1109/ASCC.2017.8287227
5. Bartolini, G., Ferrara, A.: Multi-input sliding mode control of a class of uncertain nonlinear systems. IEEE Trans. Autom. Control **41**(11), 1662–1666 (1996). https://doi.org/10.1109/9.544004
6. Utkin, V., Guldner, J., Shijun, M.: Sliding Mode Control in Electro-Mechanical Systems. CRC Press (2009)
7. Plestan, F., Shtessel, Y., Brégeault, V., Poznyak, A.: New methodologies for adaptive sliding mode control. Int. J. Control **83**(9), 1907–1919 (2010). https://doi.org/10.1080/00207179.2010.501385
8. Plestan, F., Bregeault, V., Glumineau, A., Shtessel, Y., Moulay, E.: Advances in high order and adaptive sliding mode control–theory and applications. In: Sliding Modes after the First Decade of the 21st Century, pp. 465–492. Springer (2011)
9. Bhat, S., Bernstein, D.: Finite-time stability of continuous autonomous systems. SIAM J. Control Optim. **38**(3), 751–766 (2000). https://doi.org/10.1137/S0363012997321358
10. Mukherjee, J., Roy, S., Kar, I.N., Mukherjee, S.: A double-layered artificial delay-based approach for maneuvering control of planar snake robots. J. Dyn. Syst. Meas. Control **141**(4) (2018). https://doi.org/10.1115/1.4042033
11. Mohammadi, A., Rezapour, E., Maggiore, M., Pettersen, K.Y.: Maneuvering control of planar snake robots using virtual holonomic constraints. IEEE Trans. Control Syst. Technol. **24**(3), 884–899 (2016). http://orcid.org/10.1109/TCST.2015.2467208

Chapter 3
Time-Delayed Control for Planar Snake Robots

Abstract The implementation of SMC-based techniques in Chap. 2 strictly required the uncertainties in the system to be bounded [1, 2]. However, considering these state-dependent uncertainties to be bounded can be regarded as a conservative assumption for practical scenarios. Furthermore, these approaches depend upon the viscous friction model to compute the nominal friction forces which is essential for determining the control effort. For situations where this analytical friction model doesn't hold, tracking accuracy would eventually be further compromised. Hence, to relax such assumptions and circumvent the dependence on the friction model, an artificial-delay-based methodology has been proposed in this chapter to solve the tracking problem for a planar snake robot. *Time-Delayed Control* (TDC) is a robust control method applied to solve the trajectory tracking problem for systems with model as well as parametric uncertainties and disturbances without assuming them to be bounded, utilizing input and output measurements from a delayed time. The implementation of TDC is based on the assumption that the uncertainties should vary slowly over time. *As practical snake robots are known to travel slowly through their surroundings, the uncertainty in the ground condition or the frictional forces, do evolve at a slow rate making TDC a natural and fitting choice to accomplish satisfactory tracking performance for a snake robot.*

This approach decouples the dynamics of a system using a suitable inertia matrix and combines the entire unknown part of the system dynamics into a single function. Thereafter, the unknown function is estimated from the time-delayed measurements of the control input and output [3, 4]. Subsequently, its effectiveness and simplicity have led its application to multitutde of robotics applications such as manipulators [5], aerial vehicle [6], wheeled mobile robots [7], humanoids [8], missile-interceptors [9, 10], reusable launch vehicle [11, 12], etc. From the various literature cited above, it can be observed that TDC or all the TDC-based control frameworks have been particularly used for fully actuated systems or underactuated systems. However, the snake robot, on the other hand, is an overactuated system with $(n-1)$ control inputs (τ) and 3 output space coordinates (p_x, p_y, θ_n) requiring a novel formulation to implement TDC. The contribution of this chapter is the first work toward this direction, where a Double-Layered Time-Delayed control approach is proposed for a planar snake robot. The double layering refers to the simultaneous use of TDC strategy for the body-shape control through VHCs as well as for the head-angle and velocity tracking. The proposed method derives its nomenclature from the *time-delayed estimation* (TDE) which utilizes the input and output measurements from the previous sample instant to estimate the uncertainties and state-dependent terms as an unified function. The proposed approach doesn't require any knowledge regarding the uncertainty bounds in the friction coefficients like [1]. Furthermore, it doesn't impose any limitation regarding the uncertainties to be bounded by a constant as is assumed in [1, 2]. Since boundedness of the estimation error in the individual TDC layers do not guarantee overall system stability, a common Lyapunov function-based stability analysis is proposed to demonstrate system stability. While the TDC requires small time delay for ensuring tracking accuracy [7, 8], for all practical purposes the delay is governed by the hardware specifications and is tuned to the sampling interval of the system. The stability analysis yields a tuning procedure for the controller gains where the designed gain improves tracking accuracy of the TDC, for a fixed choice of sampling interval. In this chapter, a dual-layer TDC approach has been adopted to maneuver a planar snake robot toward a desired head angle and velocity [13]. A double layer TDC framework is proposed for planar snake robot, which represents a class of overactuated system. While the outer layer TDC helps to attain the desired body shape, the inner layer TDC performs the tracking of head angle and velocity. Analysis using a combined Lyapunov function based on both layers ensures overall system stability. Moreover, the analytical study reveals how a designer can extract better tracking accuracy from TDC for a fixed sampling interval. Extensive simulations to demonstrate the stability and comparison with adaptive-sliding-mode-based method [2] establishes the efficacy of the proposed controller.

The rest of the chapter is arranged as Sect. 3.1 presents the whole TDC-based control description by detailing the outer TDC followed by the inner TDC; Sect. 3.2 presents the Lyapunov stability analysis; The simulation results and discussion are given in Sect. 3.3; Finally the chapter is summarized in Sect. 3.4.

3.1 Control Description

Generation of the desired undulations in the snake body consists of two interconnected stages: imposition of a particular gait function on the system utilizing a body-shape control and determination of the parameters of the gait function and body-shape control for head-angle and velocity tracking. To achieve these two objective simultaneously, a double-layered TDC approach is presented; while the outer layer TDC helps to achieve the desired body shape, the inner layer TDC is utilized for head-angle and velocity tracking. The body-shape control (BSC) has been designated as the *Outer loop* TDC, as this algorithm results into the actual torque control τ to be applied to the robot actuators. The head-angle and velocity control that has been assigned to be the *Inner loop* TDC yields the pseudo-controls u_{ϕ_0} and u_λ along with

3.1 Control Description

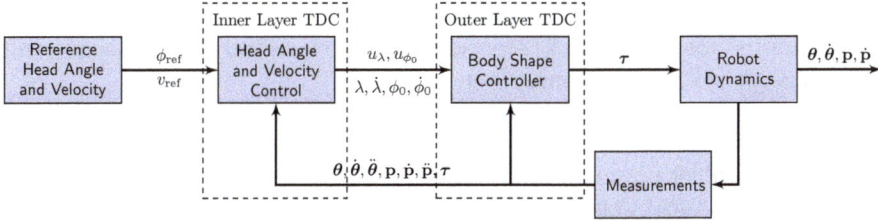

Fig. 3.1 Block diagram of control law for TDC

the dynamic parameters of the gait function and their higher order derivatives. These variables don't enter the robot actuators directly, rather influence the robot motion through the torque input emerging from the outer loop. A block diagram of the overall control scheme is shown in Fig. 3.1.

3.1.1 Outer Layer Time-Delayed Control

This subsection deals with the body-shape control where the relative joint angles are being tracked to the serpenoid gait function through VHCs. The VHC dynamics for the snake robot described through (1.8) can be expressed as

$$\ddot{\mathbf{h}} = \mathbf{D}\mathbf{M}^{-1}\left(\mathbf{W}\dot{\theta}^2 + l\mathbf{S}\mathbf{C}_\theta^T \mathbf{f}_R(\theta, \dot{\theta}, \dot{\mathbf{p}})\right) - \boldsymbol{\Phi}''(\lambda)\dot{\lambda}^2 - \boldsymbol{\Phi}'(\lambda)\ddot{\lambda} - \mathbf{b}_1\ddot{\phi}_0 + \mathbf{D}\mathbf{M}^{-1}\mathbf{D}^T \boldsymbol{\tau}. \tag{3.1}$$

Notably, the VHC system (3.1) contains uncertain terms due to the friction force vector $\mathbf{f}_R(\theta, \dot{\theta}, \dot{\mathbf{p}})$. In the previous chapter, the VHCs were bounded by utilizing a high state feedback gain due to bounded uncertainties. *Assuming these state-dependent uncertainties to be bounded can be regarded as a conservative approach. Also, it is difficult to ascertain the upper bound of such uncertainties. Hence, the VHC system also requires a dedicated robust controller for tracking the serpenoid gait.* This is the primary reason behind the employment of time-delayed controller for the designated outer loop. The VHC system (3.1) can be rearranged as

$$\mathbf{M}_h \ddot{\mathbf{h}} = \mathbf{f}_h + \boldsymbol{\tau} + \mathbf{f}_\phi, \tag{3.2}$$

where

$$\mathbf{M}_h = (\mathbf{D}\mathbf{M}^{-1}\mathbf{D}^T)^{-1},$$
$$\mathbf{f}_h = (\mathbf{D}\mathbf{M}^{-1}\mathbf{D}^T)^{-1}\mathbf{D}\mathbf{M}^{-1}\left(\mathbf{W}\dot{\theta}^2 + l\mathbf{S}\mathbf{C}_\theta^T \mathbf{f}_R(\theta, \dot{\theta}, \dot{\mathbf{p}})\right),$$
$$\mathbf{f}_\phi = (\mathbf{D}\mathbf{M}^{-1}\mathbf{D}^T)^{-1}\left(-\boldsymbol{\Phi}''(\lambda)\dot{\lambda}^2 - \boldsymbol{\Phi}'(\lambda)\ddot{\lambda} - \mathbf{b}_1\ddot{\phi}_0\right).$$

The matrix $\mathbf{M}_h$ can be identified as the inertia matrix, $\mathbf{f}_h$ contains the terms related to the Coriolis' force whereas $\mathbf{f}_\phi$ represents the terms related to the dynamic gait parameters contributed by the inner loop.

Remark: The artificial-delay-based method, TDC, is utilized to achieve a desired body shape without any knowledge of friction force vector $\mathbf{f}_R$. The TDC-based methodology neither presumes the uncertainties to be bounded by a constant, nor it requires the constant to be known as is assumed in Chap. 2. Rather, it estimates the unknown parts of the dynamics using the time-delayed approach. It is of prime importance that the estimation error should be bounded for the controller to work efficiently. For this reason, the inertia matrix of the system (3.2) is modified and the corresponding system dynamics can be presented as

$$\bar{\mathbf{M}}_h \ddot{\mathbf{h}} = \bar{\mathbf{f}}_h + \boldsymbol{\tau} + \mathbf{f}_\phi, \tag{3.3}$$

where $\bar{\mathbf{M}}_h$ is a user-defined positive definite and constant matrix and $\bar{\mathbf{f}}_h = \mathbf{f}_h + (\bar{\mathbf{M}}_h - \mathbf{M}_h)\ddot{\mathbf{h}}$. The choice of $\bar{\mathbf{M}}_h$ is critically related to the boundedness of the estimation error, to be discussed later. The control input can be designed using the inverse dynamics of (3.3) as

$$\boldsymbol{\tau} = \bar{\mathbf{M}}_h \bar{\mathbf{u}}_h - \hat{\mathbf{f}}_h - \mathbf{f}_\phi, \tag{3.4}$$

where $\hat{\mathbf{f}}_h$ is the nominal value of $\bar{\mathbf{f}}_h$ and $\bar{\mathbf{u}}_h$ is chosen as

$$\bar{\mathbf{u}}_h = -\mathbf{K}_P \mathbf{h} - \mathbf{K}_D \dot{\mathbf{h}}, \tag{3.5}$$

where $\mathbf{K}_P$ and $\mathbf{K}_D$ are positive definite matrices of appropriate dimensions. Note that, since $\mathbf{f}_R$ is considered to be unknown, the knowledge of $\hat{\mathbf{f}}_h$ is not available for computing the control input. Hence, utilizing the time-delayed estimation (TDE) process [3, 4] and considering the uncertainties to be slowly varying over the sampling interval, $\hat{\mathbf{f}}_h$ is estimated from the past input–output data. The dynamics at the past instance is utilized to estimate the unknown part $\hat{\mathbf{f}}_h(t - \gamma)$ and is used in the present time instance assuming it to be same. This approach can analytically expressed as

$$\hat{\mathbf{f}}_h \approx \bar{\mathbf{f}}_h(t - \gamma) = \bar{\mathbf{M}}_h \ddot{\mathbf{h}}(t - \gamma) - \boldsymbol{\tau}(t - \gamma) - \mathbf{f}_\phi(t - \gamma), \tag{3.6}$$

where $\gamma > 0$ is a fixed delay time and it is to be selected as the sampling interval for the TDE process [8]. Here $\boldsymbol{\tau}(t - \gamma)$ and $\ddot{\mathbf{h}}(t - \gamma)$ denote the immediate past information for $\boldsymbol{\tau}$ and $\ddot{\mathbf{h}}$, respectively. Using the control law (3.4) in the system dynamics (3.3) yields the closed-loop system as

$$\ddot{\mathbf{h}} + \mathbf{K}_P \mathbf{h} + \mathbf{K}_D \dot{\mathbf{h}} = \bar{\mathbf{M}}_h^{-1} \tilde{\mathbf{f}}_h, \tag{3.7}$$

3.1 Control Description

where $\mathbf{M}_h^{-1}\tilde{\mathbf{f}}_h = \mathbf{M}_h^{-1}(\bar{\mathbf{f}}_h - \hat{\mathbf{f}}_h) = \boldsymbol{\xi}$. The closed-loop system (3.7) can be expressed in a first-order form as

$$\dot{\bar{\mathbf{h}}} = \bar{\mathbf{K}}\bar{\mathbf{h}} + \bar{\boldsymbol{\xi}}, \tag{3.8}$$

where

$$\bar{\mathbf{h}} = \begin{bmatrix} \mathbf{h} \\ \dot{\mathbf{h}} \end{bmatrix}, \bar{\boldsymbol{\xi}} = \begin{bmatrix} \mathbf{0}_{n-1} \\ \boldsymbol{\xi} \end{bmatrix}, \bar{\mathbf{K}} = \begin{bmatrix} \mathbf{0}_{(n-1)\times(n-1)} & \mathbf{I}_{n-1} \\ -\mathbf{K}_P & -\mathbf{K}_D \end{bmatrix}.$$

Here, the vector $\boldsymbol{\xi}$ is termed as the *Bounded estimation error* or TDE error. Noticeably, the boundedness of $\boldsymbol{\xi}$ and $\hat{\boldsymbol{\xi}}$ is same and will be used in conjunction. The TDE error is found to remain bounded if $\bar{\mathbf{M}}_h$ is selected such that [3, 4, 8]

$$\left\| \mathbf{I}_{n-1} - \mathbf{M}_h^{-1}\bar{\mathbf{M}}_h \right\| < 1. \tag{3.9}$$

The proof of this boundedness condition is given in Appendix A. Notably, the boundedness of TDE error is not related to the boundedness of the uncertainty. This condition can hold true even in the presence of unbounded uncertainties.

Remark: It is noteworthy that the original system is delay free. However, an artificial delay γ is inserted into the system through the TDE process (3.6) to avoid any explicit requirement of the knowledge for the unknown system dynamics term $\mathbf{f}_R$; accordingly, the overall control framework is called time-delayed control.

Since $\mathbf{K}_P, \mathbf{K}_D$ are designed to be positive definite matrices, $\bar{\mathbf{K}}$ emerges to be a Hurwitz matrix. Hence, by virtue of the boundedness condition (3.9), the system dynamics (3.8) also remains bounded. Finally, the total control input for the outer layer TDC can be obtained by combining (3.4), (3.5) and (3.6) as [13]

$$\boldsymbol{\tau} = \boldsymbol{\tau}(t-\gamma) + \mathbf{f}_\phi(t-\gamma) - \bar{\mathbf{M}}_h \ddot{\mathbf{h}}(t-\gamma) - \mathbf{f}_\phi - \bar{\mathbf{M}}_h(\mathbf{K}_P \mathbf{h} + \mathbf{K}_D \dot{\mathbf{h}}). \tag{3.10}$$

The boundedness of the VHC $\mathbf{h}$ is ensured by the control law (3.10) acting on system (3.2) to be shown in Sect. 3.2. Once the VHCs are bounded about zero, i.e. $||\mathbf{h}|| \le \zeta$ for $\zeta > 0$, the system is said to be on the constraint manifold and the corresponding reduced-order system is said to be the constraint system. This constraint system contains the head-angle dynamics and positional dynamics of the robot CM. Tracking a particular head-angle and velocity through TDC approach on the constraint system yields the appropriate values of the gait function parameters $(\lambda, \dot{\lambda}, \phi_0, \dot{\phi}_0)$ and pseudo-inputs u_{ϕ_0}, u_λ, as discussed in the subsequent sections.

3.1.2 Inner Layer Time-Delayed Control

When the VHCs are bounded about zero, the system is said to be on the constraint manifold and the corresponding reduced system is called the constraint system.

Hence, the constraint system on the VHC manifold is considered to propose another layer of time-delayed control assigned to be *Inner Loop* TDC. The constraint system can be expressed as

$$\ddot{\theta}_n = \psi_1 + \psi_2 u_\lambda + \psi_3 u_{\phi_0}, \quad (3.11a)$$

$$\ddot{\mathbf{p}} = \bar{\mathbf{f}}_R(\theta, \dot{\theta}, \dot{\mathbf{p}}), \quad (3.11b)$$

where

$$\psi_1 = -\frac{\mathbf{e}^T \mathbf{MH}\mathbf{\Phi}''(\lambda)\dot{\lambda}^2}{\mathbf{e}^T \mathbf{Me}} - \frac{1}{\mathbf{e}^T \mathbf{Me}} \mathbf{e}^T \left(\mathbf{W}\dot{\theta}^2 - l\mathbf{SC}_\theta^T \mathbf{f}_R\right),$$

$$\bar{\mathbf{f}}_R = \frac{1}{nm} \mathbf{E}^T \mathbf{f}_R(\theta, \dot{\theta}, \dot{\mathbf{p}}),$$

$$\psi_2 = -\frac{\mathbf{e}^T \mathbf{MH}\mathbf{\Phi}'(\lambda)}{\mathbf{e}^T \mathbf{Me}},$$

$$\psi_3 = -\frac{\mathbf{e}^T \mathbf{MHb}_1}{\mathbf{e}^T \mathbf{Me}}.$$

It is to note here, that ψ_1 and $\bar{\mathbf{f}}_R$ are functions of the unknown friction forces. Whereas ψ_2 and ψ_3 being function of the inertia matrix are completely known.

Remark: For the previously presented control strategies in Chaps. 1 and 2, a viscous friction model has been assumed as proposed in [14]. The friction model has been an essential part for both the proposed control laws. But, the friction forces in practical conditions doesn't follow such equations and can be far away from such an analytical model. On the other hand, the TDC approach doesn't require any predefined friction model for implementation. Rather it estimates the forces using the input and output data from the previous instant. For this reason, the proposed TDC approach can be thought to be better suited for practical applications.

The global CM velocities are transformed from the global frame to the head-link body frame of the using transformation

$$v_t = \mathbf{u}_{\theta_n} \dot{\mathbf{p}},$$
$$v_n = \mathbf{v}_{\theta_n} \dot{\mathbf{p}},$$

where

$$\mathbf{u}_{\theta_n} = \begin{bmatrix} \cos\theta_n \\ \sin\theta_n \end{bmatrix}, \quad \mathbf{v}_{\theta_n} = \begin{bmatrix} -\sin\theta_n \\ \cos\theta_n \end{bmatrix}.$$

Henceforth, the control objectives are

$$\theta_n \to \theta_{\text{ref}} \quad ; \quad v_t \to v_{\text{ref}}.$$

3.1 Control Description

where θ_{ref} and v_{ref} are the reference head-angle and velocity, respectively. The head-angle and body-frame velocity error dynamics along with the compensators can be expressed as

$$\ddot{\tilde{\theta}}_n = -\ddot{\theta}_{\text{ref}} + \psi_2 u_\lambda + \psi_3 u_{\phi_0} + \tilde{\psi}_{\theta_n}, \tag{3.12a}$$

$$\dot{\tilde{v}}_t = f_2(\bar{\mathbf{x}}) - \dot{v}_{\text{ref}} + \tilde{\psi}_{v_t}, \tag{3.12b}$$

$$\dot{v}_n = f_3(\bar{\mathbf{x}}) + \tilde{\psi}_{v_n}, \tag{3.12c}$$

$$\ddot{\lambda} = u_\lambda, \tag{3.12d}$$

$$\ddot{\phi}_0 = u_{\phi_0}, \tag{3.12e}$$

where

$$\tilde{\psi}_{\theta_n} = \psi_1,$$
$$f_2(\bar{\mathbf{x}}) = +\mathbf{v}_{\theta_n}\dot{\mathbf{p}}\,\dot{\theta}_n,$$
$$f_3(\bar{\mathbf{x}}) = -\mathbf{u}_{\theta_n}\dot{\mathbf{p}}\,\dot{\theta}_n,$$
$$\tilde{\psi}_{v_t} = \mathbf{u}_{\theta_n}\bar{\mathbf{f}}_R(\boldsymbol{\theta},\dot{\boldsymbol{\theta}},\dot{\mathbf{p}}),$$
$$\tilde{\psi}_{v_n} = \mathbf{v}_{\theta_n}\bar{\mathbf{f}}_R(\boldsymbol{\theta},\dot{\boldsymbol{\theta}},\dot{\mathbf{p}}).$$

The head-angle error and tangential-velocity errors are defined as

$$\tilde{\theta}_n = \theta_n - \theta_{\text{ref}}, \quad \tilde{v}_t = v_t - v_{\text{ref}},$$

where θ_{ref} and v_{ref} are the reference head angle and the reference tangential velocity. The state vector for the constraint system (3.12) can be expressed as

$$\bar{\mathbf{x}} = [\tilde{\theta}_n, \dot{\tilde{\theta}}_n, \tilde{v}_t, v_n, \lambda, \dot{\lambda}, \phi_0, \dot{\phi}_0]^T \in \mathbb{R}^8.$$

With inputs u_λ and u_{ϕ_0}, the constraint system (3.12) is underactuated and thus requires output function of the same order as the input vector for control design. Therefore, the output functions for designing a TDC law are chosen to be

$$\boldsymbol{\sigma}(\bar{\mathbf{x}}) = \begin{bmatrix} \sigma_1(\bar{\mathbf{x}}) \\ \sigma_2(\bar{\mathbf{x}}) \end{bmatrix} = \begin{bmatrix} \dot{\tilde{\theta}}_n + K_n\tilde{\theta}_n \\ \dot{\lambda} + K_v\tilde{v}_t \end{bmatrix}. \tag{3.13}$$

For a system without uncertainties, the convergence of the output function (3.13) to the stable equilibrium origin ensures

$$\dot{\tilde{\theta}}_n = -K_n\tilde{\theta}_n,$$
$$\tilde{v}_t = \frac{\dot{\lambda}}{K_v}.$$

This affirms exponential convergence of the head-angle error $\tilde{\theta}_n$ to zero for a positive gain K_n and boundedness of the velocity error $\tilde{v}_t$ to a low value for a sufficiently large K_v. This is the reason behind the choice of such output function. However, stability analysis of the output function in the presence of uncertainties have been discussed in the Sect. 3.2.

First-order derivative with respect to time and consequent usage of the system dynamics (3.12a) and (3.12b) yields the output system, which can be expressed in a vector form as

$$\dot{\sigma} = \mathbf{f}_\sigma + \mathbf{g}_\sigma \mathbf{u}_\sigma + \mathbf{d}_\sigma, \tag{3.14}$$

where

$$\mathbf{f}_\sigma = \begin{bmatrix} (\dot{\hat{\psi}}_1 - \ddot{\theta}_{\text{ref}} + K_n \dot{\tilde{\theta}}_n) \\ K_v(f_2 - \dot{v}_{\text{ref}}) \end{bmatrix} \in \mathbb{R}^2, \; \mathbf{g}_\sigma = \begin{bmatrix} \psi_2 & \psi_3 \\ 1 & 0 \end{bmatrix} \in \mathbb{R}^{2\times 2}$$

$$\mathbf{u}_\sigma = \begin{bmatrix} u_\lambda \\ u_{\phi_0} \end{bmatrix} \in \mathbb{R}^2, \; \mathbf{d}_\sigma = \begin{bmatrix} \tilde{\psi}_{\theta_n} \\ \tilde{\psi}_{v_t} \end{bmatrix} \in \mathbb{R}^2.$$

The nominal part of the system vector is presented as $\mathbf{f}_\sigma$, $\mathbf{g}_\sigma$ is the input matrix, $\mathbf{u}_\sigma$ is the input vector and $\mathbf{d}_\sigma$ is the vector representing the unknown disturbances and uncertainties. In this framework, it is essential that the inverse of the input matrix $\mathbf{g}_\sigma$ should exist which have been affirmed through Theorem 2.1. Multiplying both sides of (3.14) with the inverse of the input matrix generates

$$\mathbf{g}_1 \dot{\sigma} = \mathbf{g}_1 \mathbf{f}_\sigma + \mathbf{u}_\sigma + \mathbf{g}_1 \mathbf{d}_\sigma,$$
$$\Rightarrow \mathbf{g}_1 \dot{\sigma} = \mathbf{f}_1 + \mathbf{u}_\sigma. \tag{3.15}$$

where $\mathbf{g}_1 = \mathbf{g}_\sigma^{-1}$. Similar to Sect. 3.1.1, for controller design, the inertia matrix of (3.15) is modified and expressed as

$$\bar{\mathbf{g}}_1 \dot{\sigma} = \bar{\mathbf{f}}_1 + \mathbf{u}_\sigma, \tag{3.16}$$

where $\bar{\mathbf{g}}_1$ is a user-defined positive definite and constant matrix of appropriate dimension and

$$\bar{\mathbf{f}}_1 = \mathbf{g}_1 \mathbf{f}_\sigma - (\mathbf{g}_1 - \bar{\mathbf{g}}_1)\dot{\sigma} + \mathbf{g}_1 \mathbf{d}_\sigma.$$

It is to note here that the matrix $\bar{\mathbf{g}}_1$ is chosen so as to bound the TDE error to be explained later. The control input $\mathbf{u}_\sigma$ is designed as

$$\mathbf{u}_\sigma = \bar{\mathbf{g}}_1 \bar{\mathbf{u}} - \hat{\bar{\mathbf{f}}}_1, \tag{3.17}$$

3.1 Control Description

where $\hat{\mathbf{f}}_1$ is the nominal value of $\bar{\mathbf{f}}_1$ and $\bar{\mathbf{u}}$ is an auxiliary control input designed as

$$\bar{\mathbf{u}} = -\mathbf{K}\boldsymbol{\sigma}, \tag{3.18}$$

where $\mathbf{K}$ is a positive definite gain matrix. Since $\mathbf{d}_\sigma$ is completely unknown to the user, $\hat{\mathbf{f}}_1(t)$ is designed based on the TDE method using the input–output data of the immediate past instant in the following way

$$\hat{\mathbf{f}}_1(t) \approx \bar{\mathbf{f}}_1(t-\gamma) = \bar{\mathbf{g}}_1 \dot{\boldsymbol{\sigma}}(t-\gamma) - \mathbf{u}_\sigma(t-\gamma). \tag{3.19}$$

Combining (3.17), (3.18) and TDE (3.19) yields the total control input for the inner layer TDC as [13]

$$\mathbf{u}_\sigma(t) = \mathbf{u}_\sigma(t-\gamma) - \bar{\mathbf{g}}_1 \mathbf{K}\boldsymbol{\sigma}(t) - \bar{\mathbf{g}}_1 \dot{\boldsymbol{\sigma}}(t-\gamma). \tag{3.20}$$

The closed-loop system is obtained by applying the control law (3.20) to the output system (3.16) presented as

$$\dot{\boldsymbol{\sigma}} + \mathbf{K}\boldsymbol{\sigma} = \boldsymbol{\epsilon}, \tag{3.21}$$

where $\boldsymbol{\epsilon} = \bar{\mathbf{g}}_1^{-1}(\bar{\mathbf{f}}_1 - \hat{\mathbf{f}}_1)$ denotes the TDE error for the inner layer TDC and it remains bounded if $\bar{\mathbf{g}}_1$ is selected in a way such that the following condition is satisfied:

$$\|(\mathbf{I}_2 - \mathbf{g}_1^{-1}\bar{\mathbf{g}}_1)\| < 1. \tag{3.22}$$

The proof of the condition (3.22) is provided in the Appendix B.

The following section presents the stability analysis using a Lyapunov function and considering the inner loop and the outer loop together.

3.2 Stability Analysis

The conditions (3.9) and (3.22) ensure the individual boundedness of the outer and inner layer TDC for body-shape control, head-angle and velocity tracking, respectively. However, their individual boundedness does not guarantee stable performance of the overall snake robot about the equilibrium point defined as $\{\dot{\bar{\mathbf{h}}} = 0 | \dot{\boldsymbol{\sigma}} = 0\}$. This section presents the stability analysis of the overall system about the defined operating point using the following common Lyapunov function while taking $\boldsymbol{\Xi} = [\boldsymbol{\sigma}^T \ \bar{\mathbf{h}}^T]^T$

$$V(\boldsymbol{\Xi}) = \frac{1}{2}\boldsymbol{\sigma}^T\boldsymbol{\sigma} + \frac{1}{2}\bar{\mathbf{h}}^T\boldsymbol{\Gamma}\bar{\mathbf{h}}, \tag{3.23}$$

where Γ is a positive definite matrix and the solution of the Lyapunov equation $\bar{\mathbf{K}}^T\Gamma + \Gamma\bar{\mathbf{K}} = -\Theta$ for some positive definite matrix Θ. The overall closed-loop system stability result is stated through the following theorem:

Theorem 3.1 *The system (1.8) while employing the dual-layer TDC (3.10) and (3.20) remains Uniformly Ultimately Bounded (UUB).*[1]

Proof Using (3.8)) and (3.21), the time derivative of the Lyapunov function yields

$$\dot{V} = \sigma^T \dot{\sigma} + \frac{1}{2}\bar{\mathbf{h}}^T(\bar{\mathbf{K}}^T\Gamma + \Gamma\bar{\mathbf{K}})\bar{\mathbf{h}} + \bar{\mathbf{h}}^T\Gamma\bar{\xi},$$

$$= -\sigma^T\mathbf{K}\sigma - \frac{1}{2}\bar{\mathbf{h}}^T\Theta\bar{\mathbf{h}} + \Xi^T v, \quad (3.24)$$

where $v = [\epsilon^T \ (\Gamma\bar{\xi})^T]^T$. Since Γ is a user-defined constant matrix and both ϵ and $\bar{\xi}$ remain bounded owing to the conditions (3.9)) and (3.22), one has $\|v\| \in \mathcal{L}_\infty$. Then, further simplification of (3.24) yields

$$\dot{V} \le -\lambda_{\min}(\mathbf{K})\|\sigma\|^2 - \frac{1}{2}\lambda_{\min}(\Theta)\|\bar{\mathbf{h}}\|^2 + \|v\|\|\Xi\|,$$

$$\le -\varrho_{\min}\|\Xi\|^2 + \|v\|\|\Xi\|,$$

$$= -z\|\Xi\|^2 - (\varrho_{\min} - z)\|\Xi\|^2 + \|v\|\|\Xi\|, \quad (3.25)$$

where $\varrho_{\min} = \min\{\lambda_{\min}(\mathbf{K}), \frac{1}{2}\lambda_{\min}(\Theta)\}; 0 < z < \varrho_{\min}$ is a scalar which is only used for analytical purpose. Further, the definition of V in (3.23) yields

$$V \le \varrho_{\max}(\|\sigma\|^2 + \|\bar{\mathbf{h}}\|^2) = \varrho_{\max}\|\Xi\|^2, \quad (3.26)$$

where $\varrho_{\max} = \max\{1, \|\Gamma\|\}$. Substituting the above relation in (3.25)) yields

$$\dot{V} \le -(z/\varrho_{\max})V - (\varrho_{\min} - z)\|\Xi\|^2 + \|v\|\|\Xi\|. \quad (3.27)$$

Thus, $\dot{V} \le -(z/\varrho_{\max})V < 0$ would be achieved when

$$(\varrho_{\min} - z)\|\Xi\|^2 \ge \|v\|\|\Xi\|,$$

$$\Rightarrow \|\Xi\| \ge \frac{\|v\|}{(\varrho_{\min} - z)}. \quad (3.28)$$

Hence, the overall closed-loop system would remain UUB.

Remark: From the definition of the normal velocity $v_n = \mathbf{v}_{\theta_n}\dot{\mathbf{p}}$, it can be inferred that boundedness of $\dot{\mathbf{p}}$ will ensure bounded response of v_n as well. The stability

---

[1] The solutions of system $\dot{\mathbf{x}} = \mathbf{f}(\mathbf{x}, t)$ remain Uniformly Ultimately Bounded with ultimate bound b [15], if $\exists\, a, b, c, T \in \mathbb{R}^+$ such that for $0 < a < c$ and $\|\mathbf{x}(0)\| \le a$, $\|\mathbf{x}(t)\| \le b\ \forall\, t \ge T$.

3.2 Stability Analysis

analysis proves that v_t remains bounded for all time thus affirming boundedness of $\dot{\mathbf{p}}$. This proves that v_n is also bounded for all time.

Further, it is noteworthy from (3.28) that the user can reduce the error bound by tuning the appropriate controller gains by increasing $\varrho_{\min}$ for any fixed z. It is known that reduction of sampling interval γ results in better accuracy for TDC [8]. Nevertheless, the proposed Lyapunov-based analysis also establishes an important design aspect where the designer can extract better tracking accuracy from TDC by properly tuning the gains for a fixed choice of sampling interval.

It can also be noted that the presence of the TDE error components in the numerator of (3.28), though bounded, compromises tracking accuracy. However, due to lack of space we have only concentrated on the design process of TDC for an overactuated system like snake robot. Tackling the TDE error is left as a future work. □

3.3 Simulation Environment and Results

The physical parameters considered for the robot are given in Table 2.1. The reference head angle and velocity along with the constant gait parameters are presented in Table 3.1. The matrix $\bar{\mathbf{M}}_g$ and $\bar{\mathbf{g}}_1$ satisfying condition (3.9) and (3.22) respectively have been chosen as

$$\bar{\mathbf{M}}_g = 5 \times 10^{-4} \mathbf{I}_{n-1}, \; \bar{\mathbf{g}}_1 = \begin{bmatrix} 0 & 0.6 \\ -0.6 & 0.2 \end{bmatrix}. \tag{3.29}$$

The various control gains chosen for the inner and outer loop control can be seen in Table 3.2 [13]. The performance of the proposed control scheme has been compared with the ASMC approach presented in Sect. 2.4 with the same adaptive switching gain.

Table 3.1 Reference and nominal values for TDC

Parameter	Numerical value
θ_{ref}	$-\pi/4$ rad
v_{ref}	0.05 m/s^2
α	$30\pi/180$ rad
δ	$72\pi/180$ rad

Table 3.2 Control Parameters for TDC

Parameter	Numerical value
$\mathbf{K}_P$	$10I_{n-1}$
$\mathbf{K}_D$	$10I_{n-1}$
K_n	1
K_v	14.8
$\mathbf{K}$	$30I_2$

The trajectory of the robot in the global space utilizing the proposed method and the existing method [2] is being shown in Fig. 3.2. For the proposed approach, the snake robot traverses more distance, as can be seen from the terminal phase of the trajectory, thus confirming improved velocity response which is also affirmed by Fig. 3.3. Although, for both the approaches, the steady-state head-angle converges to the reference value, in the initial phase, the proposed approach achieves the target head angle in lesser time with low initial oscillation. This fact can be confirmed through Fig. 3.4 as well. The gait frequency corresponding to the tangential velocity achieved is shown in Fig. 3.5 which shows no initial surge and steady-state oscillations within limit. In fact, the higher oscillations of the frequency results in better velocity tracking performance. The output function in Fig. 3.6 exhibits stable steady-state response with some initial oscillations for the proposed approach. The norm of VHCs in Fig. 3.7 illustrates fast convergence to the vicinity of zero for the TDC-based approach along with superior steady-state response compared to ASMC [2]. The norm of the control input $\tau(t)$ applied to track the trajectory is shown in Fig. 3.8. The exhibited evolution of the output function and the improved VHC response are the reasons behind lower steady-state control input. The initial oscillations in both output function and the VHCs result in high control input requirement in the starting phase. The estimation error from the inner and outer loop TDE have been shown in Fig. 3.9. The initial high value in the estimation error is due to unavailability of input–output data for estimation. As the input–output data becomes available to the TDE, the error reduces to a lower bound thus exemplifying the accuracy of the time-delayed estimation (Fig. 3.9).

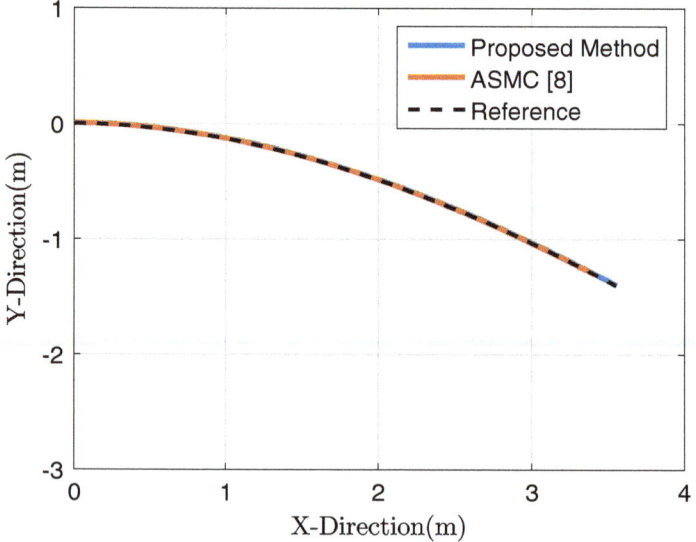

Fig. 3.2 Global trajectory of the snake robot for TDC

3.3 Simulation Environment and Results

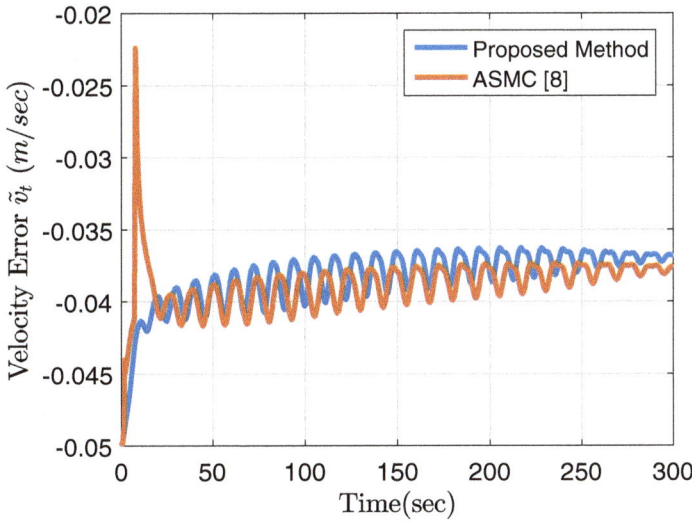

Fig. 3.3 Tangential velocity error for TDC

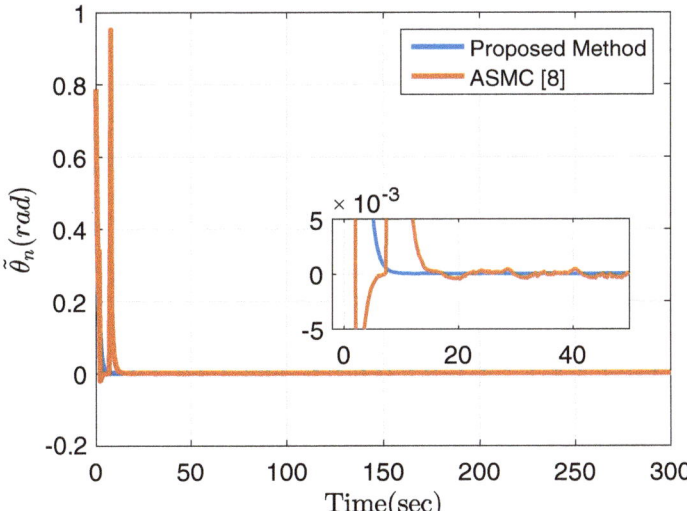

Fig. 3.4 Global head-angle error for TDC

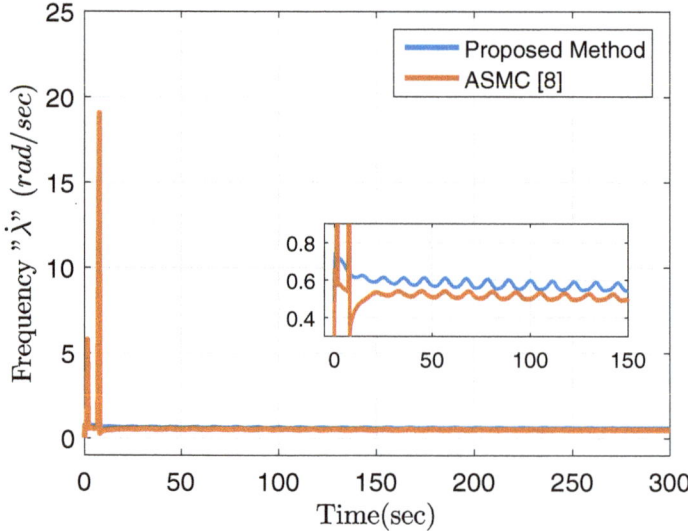

Fig. 3.5 Frequency of the gait function for TDC

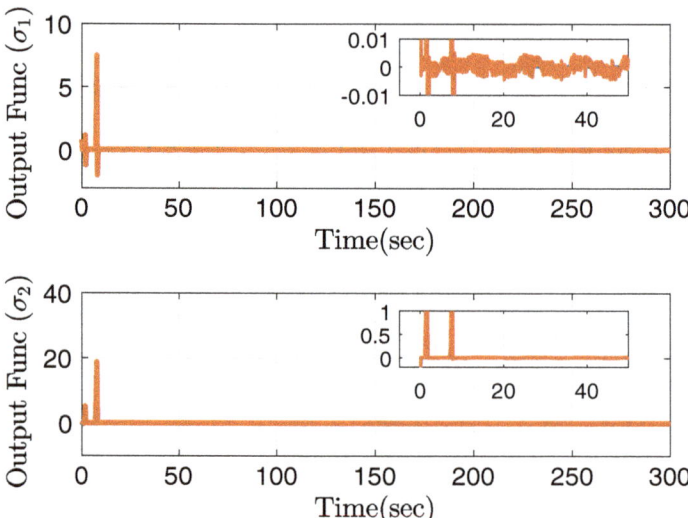

Fig. 3.6 Output function for TDC

3.3 Simulation Environment and Results

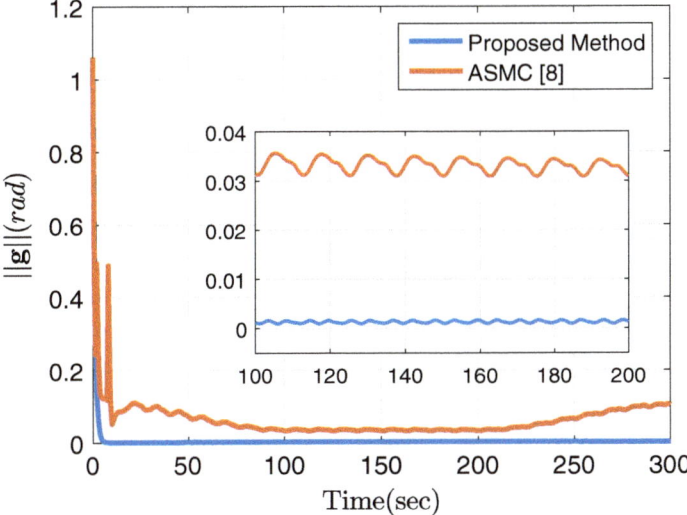

Fig. 3.7 Norm of VHCs for TDC

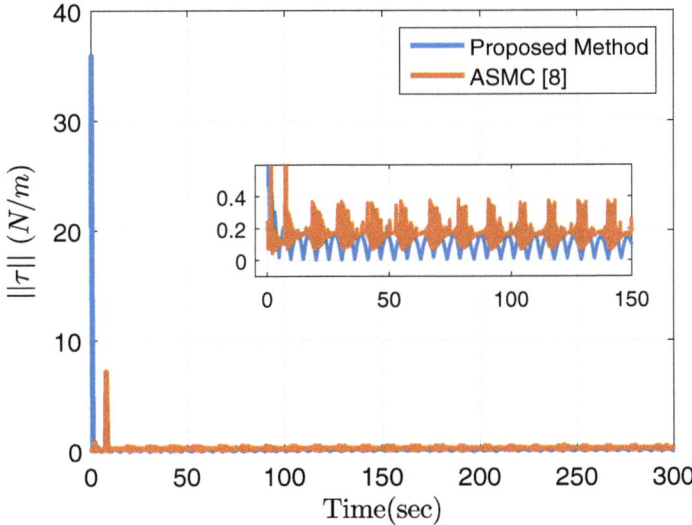

Fig. 3.8 Norm of the control effort for TDC

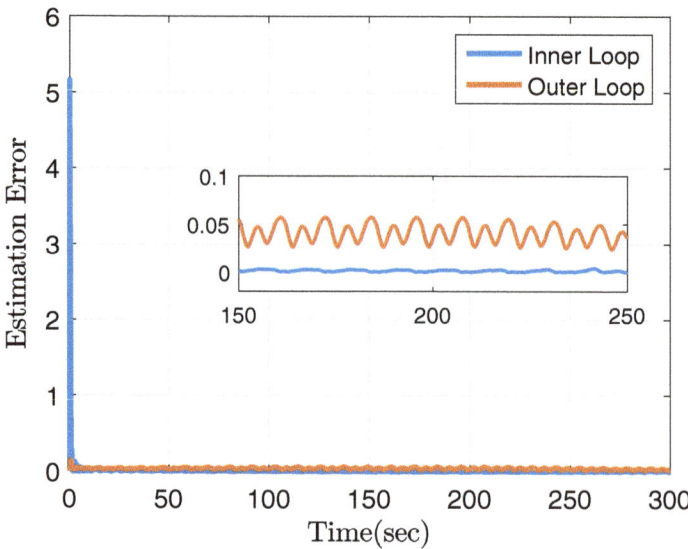

Fig. 3.9 Estimation Error for TDC

3.4 Summary

A dual-layered time-delayed control methodology for body-shape control as well as velocity and head-angle tracking has been presented in this chapter. A VHC-based approach has been adopted so that the joint angles track the pre-specified gait function. The outer layer TDC provides torque input to stabilize the VHCs in the presence of unknown uncertainties. The inner layer TDC has been applied to the reduced-order system on the VHC manifold. This layer yields the appropriate gait parameters and pseudo-inputs for the outer layer. Lyapunov-function-based analysis shows UUB stability for the closed-loop system. The proposed approach results in improved tracking performance with similar average control input. The proposed method has been found to be robust toward uncertainties simulated through time-varying friction coefficients. Future works will aim at circumventing the element of acceleration measurement from the proposed control law. Also the insertion of an additional robustness term to the proposed approach for mitigating the effect of TDE error is to be studied alongside.

References

1. Mukherjee, J., Mukherjee, S., Kar, I.N.: Sliding mode control of planar snake robot with uncertainty using virtual holonomic constraints. IEEE Robot. Autom. Lett. **2**(2), 1077–1084 (2017). http://orcid.org/10.1109/LRA.2017.2657892

References

2. Mukherjee, J., Kar, I.N., Mukherjee, S.: Adaptive sliding mode control for head-angle and velocity tracking of planar snake robot. In: 2017 11th Asian Control Conference (ASCC), pp. 537–542 (2017). https://doi.org/10.1109/ASCC.2017.8287227
3. Hsia, T.C., Gao, L.S.: Robot manipulator control using decentralized linear time-invariant time-delayed joint controllers. In: Proceedings., IEEE International Conference on Robotics and Automation, vol. 3, pp. 2070–2075 (1990). https://doi.org/10.1109/ROBOT.1990.126310
4. Youcef-Toumi, K., Ito, O.: A time delay controller for systems with unknown dynamics. In: 1988 American Control Conference, pp. 904–913 (1988). https://doi.org/10.23919/ACC.1988.4789852
5. Jin, M., Lee, J., Chang, P.H., Choi, C.: Practical nonsingular terminal sliding-mode control of robot manipulators for high-accuracy tracking control. IEEE Trans. Ind. Electron. **56**(9), 3593–3601 (2009). http://orcid.org/10.1109/TIE.2009.2024097
6. Lee, J., Yoo, C., Park, Y.S., Park, B., Lee, S.J., Gweon, D.G., Chang, P.H.: An experimental study on time delay control of actuation system of tilt rotor unmanned aerial vehicle. Mechatronics **22**(2), 184 – 194 (2012). https://doi.org/10.1016/j.mechatronics.2012.01.005. http://www.sciencedirect.com/science/article/pii/S0957415812000062
7. Roy, S., Kar, I.N., Lee, J.: Toward position-only time-delayed control for uncertain Euler-Lagrange systems: experiments on wheeled mobile robots. IEEE Robot. Autom. Lett. **2**(4), 1925–1932 (2017)
8. Roy, S., Kar, I.N., Lee, J., Jin, M.: Adaptive-robust time-delay control for a class of uncertain Euler-Lagrange systems. IEEE Trans. Ind. Electron. **64**(9), 7109–7119 (2017)
9. Banerjee, A., Mukherjee, J., un Nabi, M.: Time-energy efficient guidance strategy for a realistic 3d interceptor: an adaptive robust time-delayed control approach with input saturation. Aerosp. Sci. Technol. **104**, 106015 (2020). https://doi.org/10.1016/j.ast.2020.106015
10. Banerjee, A., Mukherjee, J., un Nabi, M., Kar, I.N.: An artificial delay based robust guidance strategy for an interceptor with input saturation. ISA Trans. (2020). https://doi.org/10.1016/j.isatra.2020.09.013
11. Sarkar, R., Mukherjee, J., Patil, D., Kar, I.N.: Re-entry trajectory tracking of reusable launch vehicle using artificial delay based robust guidance law. Advances in Space Research **67**(1), 557–570 (2021). http://orcid.org/https://doi.org/10.1016/j.asr.2020.10.006
12. Sarkar, R., Mukherjee, J., Patil, D., Kar, I.N.: Artificial time delay based adaptive robust fault tolerant control for rlv during re-entry phase. In: 2020 28th Mediterranean Conference on Control and Automation (MED), pp. 56–61 (2020). https://doi.org/10.1109/MED48518.2020.9182892
13. Mukherjee, J., Roy, S., Kar, I.N., Mukherjee, S.: A double-layered artificial delay-based approach for maneuvering control of planar snake robots. J. Dyn. Syst., Meas., Control **141**(4) (2018). https://doi.org/10.1115/1.4042033
14. Liljeback, P., Pettersen, K.Y., Stavdahl, O.: Modelling and control of obstacle-aided snake robot locomotion based on jam resolution. In: 2009 IEEE International Conference on Robotics and Automation, pp. 3807–3814 (2009). https://doi.org/10.1109/ROBOT.2009.5152273
15. Khalil, H.K., Grizzle, J.: Nonlinear Systems, vol. 3. Prentice Hall, Upper Saddle River (2002)

Chapter 4
Adaptive Robust Time-Delayed Control for Planar Snake Robots

Abstract The TDC approach presented in Chap. 3 [1] accomplished robust tracking performance of the planar snake robot in the presence of slowly varying uncertainties. Moreover, the assumption of bounded system uncertainties and the dependence on the friction model have been circumvented through the TDC methodology. However, the bounded estimation error from the TDE have an undesired effect on the overall tracking performance of the robot. This calls for a scheme to address the estimation error for achieving improved tracking accuracy without a drastic increment in the control input. For this reason, an adaptive robust controller with a novel dual-rate adaptation law has been appended with the TDC formulation to effectuate efficient tracking performance in the presence of considerable estimation error. The overall approach is thus termed as *Adaptive robust time-delayed control* (ARTDC).

ARTDC aims at extracting the advantages of an adaptive robust controller (ARC) and a TDC in a unified approach. The TDC attempts to estimate the state-dependent terms and the uncertainties as a single function, utilizing the input and output data from the previous sampling instant. To do so, an appropriately modified inertia matrix is chosen, that ensures the boundedness of the time-delayed estimation (TDE) error. This error though bounded has a pervasive effect on the controller's performance if left unattended. The feedback gain of TDC is ill-equipped to reduce this undesired effect without escalating the control input as can be seen through the stability analysis presented in the Sect. 3.2. Notably, the inclusion of a switching control law provides robustness toward the estimation error and results in to an efficient method toward improving the tracking accuracy. Furthermore, due to the absence of any information regarding the estimation error, an adaptation law has been proposed to alleviate the over–underestimation problem of the switching gain [2, 3].

The work presented in this chapter focuses on integrating *Adaptive robust* and *Time-delayed* control in a unified manner to solve the maneuvering control problem of a planar snake robot in presence of uncertainties [4]. Similar to the TDC approach, the proposed methodology has been simultaneously employed for the body-shape control through VHCs as well as for the tracking of a reference head-angle and velocity. A common Lyapunov function-based stability analysis has been presented

© The Author(s), under exclusive license to Springer Nature Switzerland AG 2021
J. Mukherjee et al., *Adaptive Robust Control for Planar Snake Robots*,
Studies in Systems, Decision and Control 363,
https://doi.org/10.1007/978-3-030-71460-4_4

in this work, as boundedness of the estimation error in the individual ARTDC layers and their individual control laws with the adaptive switching gains do not guarantee overall system stability. This work focuses on achieving better tracking accuracy with an appropriate choice of feedback gains and proper tuning of switching gains as per the conditions obtained from the stability analysis keeping the sampling interval fixed. This approach being based on the TDC based framework presented in Chap. 3 [1] doesn't require the uncertainties to be bounded as is essential for methodologies proposed in Chap. 2 [5, 6]. A double-layered ARTDC framework has been proposed for a planar snake robot where the outer layer ARTDC works to attain a desired body-shape and the inner layer ARTDC tries to track a particular head-angle and velocity. A common Lyapunov function-based stability analysis is presented that ensures the UUB stability of the overall system. The analysis yields stability conditions that guide the user towards a methodical tuning of the various gains to extract better tracking accuracy for a fixed sampling interval. The performance of the ARTDC approach has been compared with the TDC-based method through the simulation study that affirms its superiority and robustness.

The rest of the chapter is arranged as Sect. 4.1 presents the ARTDC-based control description by detailing the outer ARTDC followed by the inner ARTDC; Sect. 4.2 presents the Lyapunov stability analysis; the simulation results and discussion are given in Sect. 4.3; and finally, the chapter is summarized in Sect. 4.4.

4.1 Control Description

The total control law for the snake robot presented in this section has been segregated into two layers or loops. The outer loop is the body-shape control aimed at tracking the relative joint angles to a specified serpenoid gait function. This layer imposes the undulating body shape of the snake robot utilizing VHCs through an ATRDC approach. The velocity and heading of the robot depend upon specific parameters of the gait function. Tracking a particular velocity and head-angle has been assigned to the inner loop which employs the second layer of ARTDC. A schematic block diagram of the overall control scheme is exhibited in Fig. 4.1.

4.1.1 Outer Layer Adaptive Robust Time-Delayed Control

This section of the chapter deals with the body-shape control through an ARTDC-based approach in an attempt to track the relative joint angles to the serpenoid gait function acquiring a VHC-based approach. The VHCs (1.29) are being expressed as

$$\mathbf{h}(\lambda, \phi_0, \boldsymbol{\theta}) = \mathbf{D}\boldsymbol{\theta} - \boldsymbol{\Phi}(\lambda) - \mathbf{b}_1\phi_0, \tag{4.1}$$

4.1 Control Description

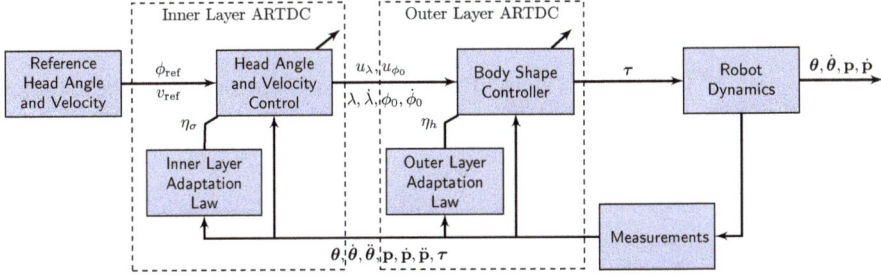

Fig. 4.1 Block diagram of control law for ARTDC

where $\boldsymbol{\Phi}(\lambda)$ represents the reference joint gait vector. The VHC system (3.2) can be obtained by differentiating (4.1) twice with respect to time as

$$\mathbf{M}_h \ddot{\mathbf{h}} = \mathbf{f}_h + \boldsymbol{\tau} + \mathbf{f}_\phi, \tag{4.2}$$

where

$$\mathbf{M}_h = (\mathbf{D}\mathbf{M}^{-1}\mathbf{D}^T)^{-1},$$
$$\mathbf{f}_h = (\mathbf{D}\mathbf{M}^{-1}\mathbf{D}^T)^{-1}\mathbf{D}\mathbf{M}^{-1}\big(\mathbf{W}\dot{\boldsymbol{\theta}}^2 + l\mathbf{S}\mathbf{C}_\theta^T \mathbf{f}_R(\boldsymbol{\theta}, \dot{\boldsymbol{\theta}}, \dot{\mathbf{p}})\big),$$
$$\mathbf{f}_\phi = (\mathbf{D}\mathbf{M}^{-1}\mathbf{D}^T)^{-1}\big(-\boldsymbol{\Phi}''(\lambda)\dot{\lambda}^2 - \boldsymbol{\Phi}'(\lambda)\ddot{\lambda} - \mathbf{b}_1\ddot{\phi}_0\big).$$

Similar to the previous chapter, the system dynamics (4.2) for TDE can be modified as

$$\bar{\mathbf{M}}_h \ddot{\mathbf{h}} = \bar{\mathbf{f}}_h + \boldsymbol{\tau} + \mathbf{f}_\phi, \tag{4.3}$$

where $\bar{\mathbf{M}}_h$ is a user-defined positive definite constant matrix to be chosen such that the TDE error is bounded. The control input is chosen as

$$\boldsymbol{\tau} = \bar{\mathbf{M}}_h \bar{\mathbf{u}}_h - \hat{\bar{\mathbf{f}}}_h - \mathbf{f}_\phi, \tag{4.4}$$

where $\hat{\bar{\mathbf{f}}}_h$ denotes the estimated value of $\bar{\mathbf{f}}_h$ obtained from the TDE and $\bar{\mathbf{u}}_h$ is designed as

$$\bar{\mathbf{u}}_h = \underbrace{-\mathbf{K}_P \mathbf{h} - \mathbf{K}_D \dot{\mathbf{h}}}_{\text{TDC}} \underbrace{-\eta_h \,\text{sat}(\mathbf{s})}_{\text{ARC}}, \tag{4.5}$$

where $\mathbf{K}_P$, $\mathbf{K}_D$ are positive definite gain matrices and η_h is a switching gain. It is to be noted here that the ARC term has been added to the TDC control law (3.5) [1] to achieve improved robustness towards the TDE error. The switching surface utilized in (4.5) can be expressed as

$$\mathbf{s} = \mathbf{B}^T \mathbf{\Gamma} \bar{\mathbf{h}}, \tag{4.6}$$

where $\bar{\mathbf{h}} = \begin{bmatrix} \mathbf{h}^T & \dot{\mathbf{h}}^T \end{bmatrix}^T$, $\mathbf{B} = \begin{bmatrix} \mathbf{0} & \mathbf{I} \end{bmatrix}^T$ and $\mathbf{\Gamma}$ is a positive definite matrix and the solution of the following Lyapunov equation

$$\bar{\mathbf{K}}^T \mathbf{\Gamma} + \mathbf{\Gamma} \bar{\mathbf{K}} = -\mathbf{\Theta}, \tag{4.7}$$

where $\mathbf{\Theta}$ is a positive definite matrix. For a small positive scalar δ_h, the saturation function employed in the switching law can be defined as

$$\text{sat}(\mathbf{s}) = \begin{cases} \frac{\mathbf{s}}{\|\mathbf{s}\|} & for \quad \|\mathbf{s}\| > \delta_h \\ \mathbf{s}/\delta_h & for \quad \|\mathbf{s}\| \leq \delta_h \end{cases}.$$

It is to be noted that $\bar{\mathbf{f}}_h$ is unavailable for controller design because the friction force vector $\mathbf{f}_R$ has been assumed to be unknown. Similar to Chap. 3 (3.6), a TDE technique has been employed utilizing the past input-output data as

$$\hat{\mathbf{f}}_h \approx \bar{\mathbf{f}}_h(t-\gamma) = \bar{\mathbf{M}}_h \ddot{\mathbf{h}}(t-\gamma) - \boldsymbol{\tau}(t-\gamma) - \mathbf{f}_\phi(t-\gamma), \tag{4.8}$$

where $\gamma > 0$ is a fixed-time delay and is selected as the sampling interval [3]. Notably, being based on the TDE, the proposed approach is independent of the analytical friction model employed in the equations of motion, elevating the applicability of this method. Substituting the control law (4.4) in the system dynamics (4.3) yields the closed loop system as

$$\ddot{\mathbf{h}} + \mathbf{K}_P \mathbf{h} + \mathbf{K}_D \dot{\mathbf{h}} + \eta_h \, \text{sat}(\mathbf{s}) = \boldsymbol{\xi}, \tag{4.9}$$

where the TDE error $\boldsymbol{\xi} = \mathbf{M}_h^{-1}(\bar{\mathbf{f}}_h - \hat{\mathbf{f}}_h)$ remains bounded for an appropriate choice of $\bar{\mathbf{M}}_h$ satisfying condition (3.9)

$$\left\| \mathbf{I}_{n-1} - \mathbf{M}_h^{-1} \bar{\mathbf{M}}_h \right\| < 1. \tag{4.10}$$

The condition (4.10) has been proven in Appendix A and implies that $\exists \eta_h^* \in \mathbb{R}^+$ such that $\|\boldsymbol{\xi}\| \geq \eta_h^*$. It is interesting to note that the original system is delay-free. The system dynamics (4.9) can be expressed in a state variable form as

$$\dot{\bar{\mathbf{h}}} = \bar{\mathbf{K}} \bar{\mathbf{h}} + \mathbf{B}(-\eta_h \, \text{sat}(\mathbf{s}) + \boldsymbol{\xi}), \tag{4.11}$$

where

$$\bar{\mathbf{K}} = \begin{bmatrix} \mathbf{0}_{(n-1) \times (n-1)} & \mathbf{I}_{n-1} \\ -\mathbf{K}_P & -\mathbf{K}_D \end{bmatrix}.$$

4.1 Control Description

The positive definiteness of $\mathbf{K}_P, \mathbf{K}_D$ make $\bar{\mathbf{K}}$ a Hurwitz matrix. Finally, the total control input for the outer layer ARTDC can be obtained by combining (4.4), (4.5) and (4.8) and can be expressed as

$$\boldsymbol{\tau} = \boldsymbol{\tau}(t-\gamma) + \mathbf{f}_\phi(t-\gamma) - \bar{\mathbf{M}}_h \ddot{\mathbf{h}}(t-\gamma) - \bar{\mathbf{M}}_h \left\{ \mathbf{K}_P \mathbf{h} + \mathbf{K}_D \dot{\mathbf{h}} + \eta_h \operatorname{sat}(\mathbf{s}) \right\} - \mathbf{f}_\phi. \tag{4.12}$$

In the previous chapter, the feedback control law of the TDC methodology is only responsible for reducing the effect of the estimation error on the tracking performance. However, the switching law injected in the ARTDC approach is dedicated to provide improved robustness against the TDE error. In a conventional robust control law, the switching gain is selected greater than an estimated upper bound of the uncertainty. Following this philosophy, one needs to select $\eta_h \geq \|\boldsymbol{\xi}\|$ to achieve desirable accuracy through the current approach. However, such selection is difficult in practice and often a conservative selection made which may also render unnecessary high switching gain and consequently higher control input. Hence, an adaptation law has been proposed in the following section to capture an appropriate switching gain that can assure robustness of the closed-loop system (4.11) without any prior knowledge of the uncertainty bound. Accordingly, the overall control framework is called ARTDC.

4.1.1.1 Adaptation Law for Switching Gain

To simultaneously maintain the robustness as well as avoid overestimation of the switching gain, a novel dual-rate adaptation law has been proposed in this section as

$$\dot{\eta}_h = \begin{cases} \eta_{h_i} \|\mathbf{s}\| & for\ \{\eta_h > 0, \mathbf{s}^T \dot{\mathbf{s}} > 0\}\ or\ \{\eta_h \leq 0\} \\ -\eta_{h_i} \|\mathbf{s}\| & for\ \{\eta_h > 0, \mathbf{s}^T \dot{\mathbf{s}} \leq 0\} \end{cases}, \tag{4.13}$$

where

$$i = \begin{cases} 1 & for\quad \|\mathbf{s}\| \leq \delta_h \\ 2 & for\quad \|\mathbf{s}\| > \delta_h \end{cases}, \text{ and } \eta_h(t_0) > 0,$$

where $\eta_{h_2} > \eta_{h_1} > 0$ are adaptation gains, δ_h is a positive scalar and t_0 is the initial time.

The adaptation law has been designed as a dual rate, i.e. two different rates of update depending upon 2-norm of the switching variable $\|\mathbf{s}\|$. A positive constant δ_h has been selected to represent the vicinity of the reference point or origin. The adaptation rate is chosen to be higher η_{h_2} for $\|\mathbf{s}\| > \delta_h$, i.e. if the switching variable is not in the vicinity of origin. Once it is found to be in the vicinity, the rate of adaptation is reduced to a lower value η_{h_1}. The decision on increasing or decreasing the gain has been relied upon the sign of the term $\bar{\mathbf{h}}^T \dot{\bar{\mathbf{h}}}$. The initial condition and the adaptive law (4.13) ensures that $\eta_h \geq 0\ \forall t \geq t_0$.

With the VHCs being bounded about zero, the system is said to be on the constrained manifold and the corresponding reduced order system is known as the constraint system. The constraint system acts as the foundation of the inner loop head-angle and velocity control. This ARTDC-based approach yields the appropriate values of the gait function parameters ($\lambda, \dot{\lambda}, \phi_0, \dot{\phi}_0$) and pseudo-inputs ($u_{\phi_0}, u_\lambda$) to track a reference head-angle and velocity, as discussed in the subsequent sections. These parameters build up the vector $\mathbf{f}_\phi$ that signifies the contribution of the inner loop control to the overall system control input (4.12).

Lemma 4.1 *For the system* (4.2) *along with TDC law* (4.12) *and switching gain adaptation law* (4.13), $\exists \eta_h^* > 0$ *such that the switching gain η_h is bounded, i.e.*

$$\eta_h(t) \leq \eta_h^*, \ \forall \ t > 0.$$

The proof to this lemma has been given in Appendix C.1.

4.1.2 Inner Layer Adaptive Robust Time-Delayed Control

This section presents the control framework for head-angle and velocity tracking through the inner layer ARTDC after achieving boundedness of the VHCs in the previous section. It is to be noted here that, the aim of this layer is to yield appropriate values of the gait parameters and the function $\mathbf{f}_\phi$ to be utilized in the outer layer control input (4.12) to achieve a desired heading and velocity. The output system on the constraint manifold (3.14) can be presented in a vector form as

$$\dot{\boldsymbol{\sigma}} = \mathbf{f}_\sigma + \mathbf{g}_\sigma \mathbf{u}_\sigma + \mathbf{d}_\sigma. \tag{4.14}$$

The nominal part of the system vector, represented as $\mathbf{f}_\sigma$ represents the known and measurable part of the output dynamics whereas the uncertainties have been clubbed into the vector $\mathbf{d}_\sigma$. The input matrix is symbolized by $\mathbf{g}_\sigma$ and the corresponding input by the vector $\mathbf{u}_\sigma$. The input matrix $\mathbf{g}_\sigma$ being invertible as has been assured in Theorem 2.1, the system in (4.14) can be rearranged as

$$\mathbf{g}_1 \dot{\boldsymbol{\sigma}} = \mathbf{g}_1 \mathbf{f}_\sigma + \mathbf{u}_\sigma + \mathbf{g}_1 \mathbf{d}_\sigma, \tag{4.15}$$

where $\mathbf{g}_1 = \mathbf{g}_\sigma^{-1}$. In coherence with ARTDC methodology, the system (4.15) is expressed as

$$\bar{\mathbf{g}}_1 \dot{\boldsymbol{\sigma}} = \bar{\mathbf{f}}_1 + \mathbf{u}_\sigma, \tag{4.16}$$

where $\bar{\mathbf{g}}_1$ is a user-defined positive definite and constant matrix of appropriate dimension and

4.1 Control Description

$$\bar{\mathbf{f}}_1 = \mathbf{g}_1 \mathbf{f}_\sigma + (\mathbf{g}_1 - \bar{\mathbf{g}}_1)\dot{\sigma} + \mathbf{g}_1 \mathbf{d}_\sigma.$$

Utilizing inverse dynamics method, the control input $\mathbf{u}_\sigma$ can be written as

$$\mathbf{u}_\sigma = \bar{\mathbf{g}}_1 \bar{\mathbf{u}} - \hat{\mathbf{f}}_1, \tag{4.17}$$

where $\hat{\mathbf{f}}_1$ is the time-delayed estimated value of $\bar{\mathbf{f}}_1$. The auxiliary control input $\bar{\mathbf{u}}$ in (4.17) is designed as

$$\bar{\mathbf{u}} = -\mathbf{K}\sigma - \eta_\sigma \text{sat}(\sigma), \tag{4.18}$$

where $\mathbf{K}$ is a positive definite gain matrix, η_σ is the switching gain. The saturated function of the system output for a small positive scalar δ_σ can be described as

$$\text{sat}(\sigma) = \begin{cases} \frac{\sigma}{\|\sigma\|} & for \quad \|\sigma\| > \delta_\sigma \\ \sigma/\delta_\sigma & for \quad \|\sigma\| \le \delta_\sigma \end{cases}.$$

As in the outer layer, the vector $\bar{\mathbf{f}}_1(t)$ is estimated through the TDE method to yield $\hat{\mathbf{f}}_1(t)$ using the input–output data of the immediate past instant illustrated as

$$\hat{\mathbf{f}}_1(t) \approx \bar{\mathbf{f}}_1(t - \gamma) = \bar{\mathbf{g}}_1 \dot{\sigma}(t - \gamma) - \mathbf{u}_\sigma(t - \gamma).$$

Notably, $\bar{\mathbf{f}}_1(t)$ contains all the uncertainties in the system and the functions of the viscous friction model. Hence, by estimating $\bar{\mathbf{f}}_1(t)$ through TDE, this approach obviates the requirement of any information of the uncertainties and circumvents the dependence on the friction model.

Substituting the control law (4.17) and auxiliary control input (4.18) into the system (4.16), the closed-loop system becomes

$$\dot{\sigma} + \mathbf{K}\sigma + \eta_\sigma \text{sat}(\sigma) = \epsilon, \tag{4.19}$$

where $\epsilon = \bar{\mathbf{g}}_1^{-1}(\bar{\mathbf{f}}_1 - \hat{\mathbf{f}}_1)$ denotes the TDE error for the inner layer ARTDC. It is desired to ensure that the TDE error ϵ remains bounded. To do so, the modified inertia matrix $\bar{\mathbf{g}}_1$ is selected in a way such that the following condition is satisfied:

$$\|(\mathbf{I}_2 - \mathbf{g}_1^{-1}\bar{\mathbf{g}}_1)\| < 1. \tag{4.20}$$

Hence, the condition (4.20) implies that $\exists \eta_\sigma^* \in \mathbb{R}^+$ such that $\|\epsilon\| \le \eta_\sigma^*$. The overall control input for the inner-layer ARTDC can be cumulatively described as

$$\mathbf{u}_\sigma(t) = \mathbf{u}_\sigma(t - \gamma) - \bar{\mathbf{g}}_1 \mathbf{K}\sigma(t) - \bar{\mathbf{g}}_1 \eta_\sigma \text{sat}(\sigma(t)) - \bar{\mathbf{g}}_1 \dot{\sigma}(t - \gamma). \tag{4.21}$$

As in the outer layer, due to the unavailability of any information regarding the estimation error, an adaptation law has been employed which subsequently alleviates the over-underestimation of the switching gain.

4.1.2.1 Adaptation Law for Switching Gain

Without any information regarding the system uncertainties, it is arduous to evaluate the estimation error ϵ. However, the TDE error being bounded, appropriate choice of the switching gain η_σ can yield desired tracking performance. Hence, to achieve robustness against the TDE error in the inner layer control without knowing its prior bound, an adaptation law for the switching gain η_σ, following similar approach like Sect. 4.1.1.1, is proposed as

$$\dot{\eta}_\sigma = \begin{cases} \eta_{\sigma_i} \|\boldsymbol{\sigma}\| & for \ \{\eta_\sigma > 0, \boldsymbol{\sigma}^T \dot{\boldsymbol{\sigma}} > 0\} \ or \ \{\eta_\sigma \leq 0\} \\ -\eta_{\sigma_i} \|\boldsymbol{\sigma}\| & for \ \{\eta_\sigma > 0, \boldsymbol{\sigma}^T \dot{\boldsymbol{\sigma}} \leq 0\} \end{cases}, \quad (4.22)$$

where

$$i = \begin{cases} 1 & for \quad \|\boldsymbol{\sigma}\| \leq \delta_\sigma \\ 2 & for \quad \|\boldsymbol{\sigma}\| > \delta_\sigma \end{cases}, \text{ and } \eta_\sigma(t_0) > 0,$$

where $\eta_{\sigma_2} > \eta_{\sigma_1} > 0$ are adaptation gains and δ_σ is a positive scalar that represents the vicinity of the origin. The philosophy behind the design of this adaptation law is same as described for the outer layer.

Lemma 4.2 *For the system* (4.14) *along with TDC law* (4.21) *and switching gain adaptation law* (4.22), $\exists \eta_\sigma^* > 0$ *such that the switching gain η_σ is bounded, i.e.*

$$\eta_\sigma(t) \leq \eta_\sigma^*, \ \forall \ t > 0.$$

The proof to this lemma has been enclosed in Appendix C.2.

4.2 Stability Analysis

4.2.1 Stability Analysis of Dual-Adaptive Robust Time-Delayed Control

The conditions (4.10) and (4.20) ensure the individual boundedness of the outer and inner layer TDE errors. However, the stability of the overall closed-loop system is of critical importance. This aspect is studied in this section using the following Lyapunov function while taking $\boldsymbol{\Xi} = [\boldsymbol{\sigma}^T \ \bar{\mathbf{h}}^T]^T$ and $\boldsymbol{\eta} = [\eta_h \ \eta_\sigma]^T$

4.2 Stability Analysis

$$V(\Xi, \eta) = V_1(\Xi) + V_2(\eta),$$
$$= \frac{1}{2}\bar{\mathbf{h}}^T \mathbf{\Gamma} \bar{\mathbf{h}} + \frac{1}{2}\sigma^T \sigma + \frac{1}{2\gamma_h}(\eta_h - \eta_h^*)^2 + \frac{1}{2\gamma_\sigma}(\eta_\sigma - \eta_\sigma^*)^2. \quad (4.23)$$

The closed-loop system stability result for the overall system is stated in the following theorem:

Theorem 4.1 *The system (1.8) employing the dual-layer ARTDC (4.12) and (4.21) with adaptation laws (4.13) and (4.22), respectively, remains UUB.*

Proof The time derivative of the Lyapunov function yields

$$\dot{V} = \frac{1}{2}\dot{\bar{\mathbf{h}}}^T \mathbf{\Gamma} \bar{\mathbf{h}} + \frac{1}{2}\bar{\mathbf{h}}^T \mathbf{\Gamma} \dot{\bar{\mathbf{h}}} + \sigma^T \dot{\sigma} + \frac{1}{\gamma_h}(\eta_h - \eta_h^*)\dot{\eta}_h + \frac{1}{\gamma_\sigma}(\eta_\sigma - \eta_\sigma^*)\dot{\eta}_\sigma.$$

Utilizing the closed-loop systems (4.11) with (4.7) and (4.19),

$$\dot{V} = -\frac{1}{2}\bar{\mathbf{h}}^T \mathbf{\Theta} \bar{\mathbf{h}} - \sigma^T \mathbf{K} \sigma + \mathbf{s}^T \xi + \sigma^T \epsilon - \eta_\sigma \sigma^T \operatorname{sat}(\sigma)$$
$$- \eta_h \mathbf{s}^T \operatorname{sat}(\mathbf{s}) + \frac{1}{\gamma_h}(\eta_h - \eta_h^*)\dot{\eta}_h + \frac{1}{\gamma_\sigma}(\eta_\sigma - \eta_\sigma^*)\dot{\eta}_\sigma,$$

$$\Rightarrow \dot{V} = -\Xi^T \mathbf{\Upsilon} \Xi + \mathbf{s}^T \xi + \sigma^T \epsilon - \eta_\sigma \|\sigma\| - \eta_h \|\mathbf{s}\|$$
$$+ \frac{1}{\gamma_h}(\eta_h - \eta_h^*)\dot{\eta}_h + \frac{1}{\gamma_\sigma}(\eta_\sigma - \eta_\sigma^*)\dot{\eta}_\sigma, \quad (4.24)$$

where

$$\mathbf{\Upsilon} = \begin{bmatrix} \frac{1}{2}\mathbf{\Theta} & 0 \\ 0 & \mathbf{K} \end{bmatrix}.$$

Case 1: $\|\mathbf{s}\| \leq \delta_h$ and $\|\sigma\| \leq \delta_\sigma$

The system states in this case are already bounded since $\|\mathbf{s}\| \leq \delta_h$ and $\|\sigma\| \leq \delta_\sigma$. Further, the gains η_h and η_σ also remain bounded as has been proved in Appendix C.1 and C.2. Hence, the boundedness for Case 1 is not shown separately to avoid repetition.

Case 2: $\left\{\{\eta_h > 0, \mathbf{s}^T \dot{\mathbf{s}} > 0\} \text{ or } \{\eta_h \leq 0\}\right\} \cup \left\{\{\eta_\sigma > 0, \sigma^T \dot{\sigma} > 0\} \text{ or } \{\eta_\sigma \leq 0\}\right\}$

$$\dot{V} = -\Xi^T \mathbf{\Upsilon} \Xi + \mathbf{s}^T \xi + \sigma^T \epsilon - \eta_\sigma \|\sigma\| - \eta_h \|\mathbf{s}\|$$
$$+ \frac{1}{\gamma_h}(\eta_h - \eta_h^*)\eta_{h_2}\|\mathbf{s}\| + \frac{1}{\gamma_\sigma}(\eta_\sigma - \eta_\sigma^*)\eta_{\sigma_2}\|\sigma\|.$$

Since $\gamma_h = \eta_{h_2}, \gamma_\sigma = \eta_{\sigma_2}$ for $\|\mathbf{s}\| > \delta_h$ and $\|\sigma\| > \delta_\sigma$, one has

$$\dot{V} \leq -\lambda_{\min}(\boldsymbol{\Upsilon}) \|\boldsymbol{\Xi}\|^2 - (\eta_h - \eta_h^*) \|\mathbf{s}\| - (\eta_\sigma - \eta_\sigma^*) \|\boldsymbol{\sigma}\|$$
$$+ (\eta_h - \eta_h^*) \|\mathbf{s}\| + (\eta_\sigma - \eta_\sigma^*) \|\boldsymbol{\sigma}\|,$$
$$\leq -\lambda_{\min}(\boldsymbol{\Upsilon}) \|\boldsymbol{\Xi}\|^2 \leq 0.$$

Thus, the above condition implies $V(t) \in \mathcal{L}_\infty \Rightarrow \boldsymbol{\sigma}, \bar{\mathbf{h}} \in \mathcal{L}_\infty$.

Case 3: $\left\{\{\eta_h > 0, \mathbf{s}^T \dot{\mathbf{s}} > 0\} \text{ or } \{\eta_h \leq 0\}\right\} \cup \left\{\eta_\sigma > 0, \boldsymbol{\sigma}^T \dot{\boldsymbol{\sigma}} \leq 0\right\}$

$$\dot{V} = -\boldsymbol{\Xi}^T \boldsymbol{\Upsilon} \boldsymbol{\Xi} + \mathbf{s}^T \boldsymbol{\xi} + \boldsymbol{\sigma}^T \boldsymbol{\epsilon} - \eta_\sigma \|\boldsymbol{\sigma}\| - \eta_h \|\mathbf{s}\|$$
$$+ \frac{1}{\gamma_h}(\eta_h - \eta_h^*)\eta_{h_2} \|\mathbf{s}\| - \frac{1}{\gamma_\sigma}(\eta_\sigma - \eta_\sigma^*)\eta_{\sigma_2} \|\boldsymbol{\sigma}\|,$$
$$\leq -\lambda_{\min}(\boldsymbol{\Upsilon}) \|\boldsymbol{\Xi}\|^2 - (\eta_h - \eta_h^*) \|\mathbf{s}\| - (\eta_\sigma - \eta_\sigma^*) \|\boldsymbol{\sigma}\|$$
$$+ (\eta_h - \eta_h^*) \|\mathbf{s}\| - (\eta_\sigma - \eta_\sigma^*) \|\boldsymbol{\sigma}\|,$$
$$\leq -\lambda_{\min}(\boldsymbol{\Upsilon}) \|\boldsymbol{\Xi}\|^2 + 2\eta_\sigma^* \|\boldsymbol{\Xi}\|.$$

Considering a scalar $0 < z < \lambda_{\min}(\boldsymbol{\Upsilon})$, the above relation can be written as

$$\dot{V} \leq -z \|\boldsymbol{\Xi}\|^2 - (\lambda_{\min}(\boldsymbol{\Upsilon}) - z) \|\boldsymbol{\Xi}\|^2 + 2\eta_\sigma^* \|\boldsymbol{\Xi}\|. \tag{4.25}$$

Further, the definition of the Lyapunov function (4.23) and the Lemma 4.2 yields

$$V \leq \varrho \|\boldsymbol{\Xi}\|^2 + \bar{\eta}, \tag{4.26}$$

where $\varrho \triangleq \max\{\|\boldsymbol{\Gamma}\|, 1\}$ and $\bar{\eta} \triangleq \frac{1}{2\gamma_h}(\bar{\eta}_h^2 + \eta_h^{*2}) + \frac{1}{2\gamma_\sigma}(\bar{\eta}_\sigma^2 + \eta_\sigma^{*2})$.
Using (4.26), the relation (4.25) can be modified as

$$\dot{V} \leq -\varrho_1 V - (\lambda_{\min}(\boldsymbol{\Upsilon}) - z) \|\boldsymbol{\Xi}\|^2 + 2\eta_\sigma^* \|\boldsymbol{\Xi}\| + \varrho_1 \bar{\eta}, \tag{4.27}$$

where $\varrho_1 \triangleq (z/\varrho)$. Hence, $\dot{V} < 0$ would be achieved when

$$(\lambda_{\min}(\boldsymbol{\Upsilon}) - z) \|\boldsymbol{\Xi}\|^2 \geq 2\eta_\sigma^* \|\boldsymbol{\Xi}\| + \varrho_1 \bar{\eta},$$

$$\|\boldsymbol{\Xi}\| \geq \frac{\eta_\sigma^*}{(\lambda_{\min}(\boldsymbol{\Upsilon}) - z)} + \sqrt{\frac{\eta_\sigma^{*2}}{(\lambda_{\min}(\boldsymbol{\Upsilon}) - z)^2} + \frac{\varrho_1 \bar{\eta}}{(\lambda_{\min}(\boldsymbol{\Upsilon}) - z)}}. \tag{4.28}$$

This affirms UUB stability for Case 3.

Case 4: $\left\{\eta_h > 0, \mathbf{s}^T \dot{\mathbf{s}} \leq 0\right\} \cup \left\{\{\eta_\sigma > 0, \boldsymbol{\sigma}^T \dot{\boldsymbol{\sigma}} > 0\} \text{ or } \{\eta_\sigma \leq 0\}\right\}$

$$\dot{V} = -\boldsymbol{\Xi}^T \boldsymbol{\Upsilon} \boldsymbol{\Xi} + \mathbf{s}^T \boldsymbol{\xi} + \boldsymbol{\sigma}^T \boldsymbol{\epsilon} - \eta_\sigma \|\boldsymbol{\sigma}\| - \eta_h \|\mathbf{s}\|$$
$$- \frac{1}{\gamma_h}(\eta_h - \eta_h^*)\eta_{h_2} \|\mathbf{s}\| + \frac{1}{\gamma_\sigma}(\eta_\sigma - \eta_\sigma^*)\eta_{\sigma_2} \|\boldsymbol{\sigma}\|,$$

4.2 Stability Analysis

$$\leq -\lambda_{\min}(\mathbf{\Upsilon}) \|\mathbf{\Xi}\|^2 - (\eta_h - \eta_h^*) \|\mathbf{s}\| - (\eta_\sigma - \eta_\sigma^*) \|\boldsymbol{\sigma}\|$$
$$- (\eta_h - \eta_h^*) \|\mathbf{s}\| + (\eta_\sigma - \eta_\sigma^*) \|\boldsymbol{\sigma}\|,$$
$$\leq -\lambda_{\min}(\mathbf{\Upsilon}) \|\mathbf{\Xi}\|^2 + 2\eta_h^* \|\mathbf{B}^T \mathbf{\Gamma}\| \|\mathbf{\Xi}\|.$$

Following the similar procedure like Case 3, the $\dot{V} < 0$ condition for Case 4 is achieved when

$$\|\mathbf{\Xi}\| \geq \frac{\eta_h^* \|\mathbf{B}^T \mathbf{\Gamma}\|}{(\lambda_{\min}(\mathbf{\Upsilon}) - z)} + \sqrt{\frac{(\eta_h^* \|\mathbf{B}^T \mathbf{\Gamma}\|)^2}{(\lambda_{\min}(\mathbf{\Upsilon}) - z)^2} + \frac{\varrho_1 \gamma}{(\lambda_{\min}(\mathbf{\Upsilon}) - z)}}. \quad (4.29)$$

This affirms UUB stability for Case 4.

Case 5: $\left\{ \eta_h > 0, \mathbf{s}^T \dot{\mathbf{s}} \leq 0 \right\} \cup \left\{ \eta_\sigma > 0, \boldsymbol{\sigma}^T \dot{\boldsymbol{\sigma}} \leq 0 \right\}$

$$\dot{V} = -\mathbf{\Xi}^T \mathbf{\Upsilon} \mathbf{\Xi} + \mathbf{s}^T \boldsymbol{\xi} + \boldsymbol{\sigma}^T \boldsymbol{\epsilon} - \eta_\sigma \|\boldsymbol{\sigma}\| - \eta_h \|\mathbf{s}\|$$
$$- \frac{1}{\gamma_h}(\eta_h - \eta_h^*)\eta_{h_2} \|\mathbf{s}\| - \frac{1}{\gamma_\sigma}(\eta_\sigma - \eta_\sigma^*)\eta_{\sigma_2} \|\boldsymbol{\sigma}\|,$$
$$\leq -\lambda_{\min}(\mathbf{\Upsilon}) \|\mathbf{\Xi}\|^2 + 2\eta_h^* \|\mathbf{B}^T \mathbf{\Gamma}\| \|\bar{\mathbf{h}}\| + 2\eta_\sigma^* \|\boldsymbol{\sigma}\|.$$

Since $\mathbf{\Xi} = [\boldsymbol{\sigma}^T \ \bar{\mathbf{g}}^T]^T$ implies $\|\mathbf{\Xi}\| \geq \|\bar{\mathbf{h}}\|$, $\|\mathbf{\Xi}\| \geq \|\boldsymbol{\sigma}\|$, one has

$$\dot{V} \leq -\lambda_{\min}(\mathbf{\Upsilon}) \|\mathbf{\Xi}\|^2 + 2\eta^* \|\mathbf{\Xi}\|, \quad (4.30)$$

where $\eta^* \triangleq \max \left\{ \eta_h^* \|\mathbf{B}^T \mathbf{\Gamma}\|, \ \eta_\sigma^* \right\}$. The condition for $\dot{V} < 0$ can be obtained as

$$\|\mathbf{\Xi}\| \geq \frac{\eta^*}{(\lambda_{\min}(\mathbf{\Upsilon}) - z)} + \sqrt{\frac{\eta^{*2}}{(\lambda_{\min}(\mathbf{\Upsilon}) - z)^2} + \frac{\varrho_1 \gamma}{(\lambda_{\min}(\mathbf{\Upsilon}) - z)}}. \quad (4.31)$$

This affirms UUB stability for Case 5. Thus, the overall closed-loop system has been proved to remain UUB. □

4.2.2 Selection of Parameters

The various control parameters $\mathbf{K}_P$, $\mathbf{K}_D$, $\mathbf{K}$, K_n, K_v, η_{h_1}, η_{h_2}, η_{σ_1}, η_{σ_2}, δ_h and δ_σ are required to be chosen appropriately to accomplish satisfactory tracking performance. Also, one needs to choose $\bar{\mathbf{M}}_h$ and $\bar{\mathbf{g}}_1$ according to the conditions (4.10) and (4.20) respectively. The control gains $\mathbf{K}_P$ and $\mathbf{K}_D$ are chosen so as to obtain a Hurwitz $\bar{\mathbf{K}}$ matrix for the second-order closed-loop system (4.9). For this reason, the matrices are selected as $\mathbf{K}_P = \omega_n^2 I_{n-1}$ and $\mathbf{K}_D = 2\zeta \omega_n I_{n-1}$, where ω_n and ζ are the desired natural frequency and damping coefficient for the nominal outer-layer closed-loop system.

Similarly, **K** is chosen as the desired rate of convergence for the nominal inner-layer closed-loop system (4.19). The sampling interval γ is required to be chosen as small as possible, although, it primarily depends upon the hardware capability. For the nominal system of the inner layer, while the high positive value of K_n ensures faster convergence of the head angle error, large value of K_v reduces velocity tracking error by increasing the gait frequency.

The positive gains η_{h_1}, η_{h_2}, η_{σ_1} and η_{σ_2}, related to the adaptation law for both the layers dictate the rate of adaptation space,i.e. the rate at which the switching gains are to change based upon the tracking error. The choice of these adaptation rates is crucial as high values can lead to faster adaptation but at the cost of high control input, whereas small adaptation rates can deteriorate the tracking performance thus jeopardizing the whole control objective. The parameters δ_h and δ_σ define the vicinity of the equilibrium point for the outer and inner layer respectively. These parameters dictate the switching of the adaptation law from a higher rate to a lower rate and vice versa. The choice of these parameters is also crucial in a way that, small values of these may lead to consistent higher adaptation rate increases the chances of high gain, whereas high values may not result in the convergence of the states to the equilibrium point. Notably, multiple possible combinations of the aforementioned parameters exist that can satisfy the control objectives. However, one needs to select these parameters according to the desired performance in the practical scenario for various applications.

4.3 Simulation Results

The various performance criteria that exemplify the different aspects of the control approach are being presented in this section as function of time. The physical parameters of the snake robot considered for simulation are proposed in Table 2.1. The modified inertia matrices according to condition (4.10) and (4.20) have chosen to be same as in the previous chapter (3.29). A stricter uncertainty regime has been considered to study the performance of the proposed control law in comparison to the TDC-based methodology. The time varying friction coefficients c_t and c_n are chosen as

$$c_t = (c_{t_2} - c_{t_1}) \operatorname{rand}(1) + c_{t_1}, \tag{4.32a}$$

$$c_n = (c_{n_2} - c_{n_1}) \operatorname{rand}(1) + c_{n_1}, \tag{4.32b}$$

where c_{t_1} and c_{t_2} are the lower and upper limit for tangential friction coefficient, respectively, and c_{n_1} and c_{n_2} are the same for coefficient along the normal direction. The above function has been designed to select a random value for the friction coefficients within a predefined range. To conserve controllability of the system through anisotropic friction, the tangential and normal coefficient bounds are chosen satisfying the condition $c_{n_1} > c_{t_2}$ as given in Table 4.1 [4].

4.3 Simulation Results

Table 4.1 Friction coefficient bound for ARTDC

Parameter	Numerical value
c_{t_1}	0.1
c_{t_2}	0.5
c_{n_1}	0.6
c_{n_2}	3

Table 4.2 Controller parameters for ARTDC

Parameter	Numerical Value
$\mathbf{K}_P$	$10I_{n-1}$
$\mathbf{K}_D$	$10I_{n-1}$
K_n	1
K_v	15.2
$\mathbf{K}$	I_2
δ_h	0.001
$\eta_h(0)$	1
η_{h_1}	10
η_{h_2}	20
δ_σ	0.01
η_{σ_1}	1
η_{σ_2}	5
$\eta_\sigma(0)$	2

The reference and nominal values have been chosen to be same as in the previous chapter presented in Table 3.1. The control gains and adaptation law related parameters tuned for the simulations are being presented in Table 4.2 [4]. The matrix $\mathbf{\Gamma}$ has been chosen as

$$\mathbf{\Gamma} = \begin{bmatrix} 5.05\ I_{n-1} & 0.25\ I_{n-1} \\ 0.25\ I_{n-1} & 0.255\ I_{n-1} \end{bmatrix}.$$

4.3.1 Discussion

The trajectory of the robot CM by virtue of the proposed approach in comparison to the TDC methodology presented in Chap. 3, has been shown in Fig. 4.2. This figure clearly shows the superior trajectory tracking performance of the proposed method. This is further attested by Figs. 4.3 and 4.4 which evidently demonstrate the improvement in the velocity tracking as well as the head-angle tracking performance

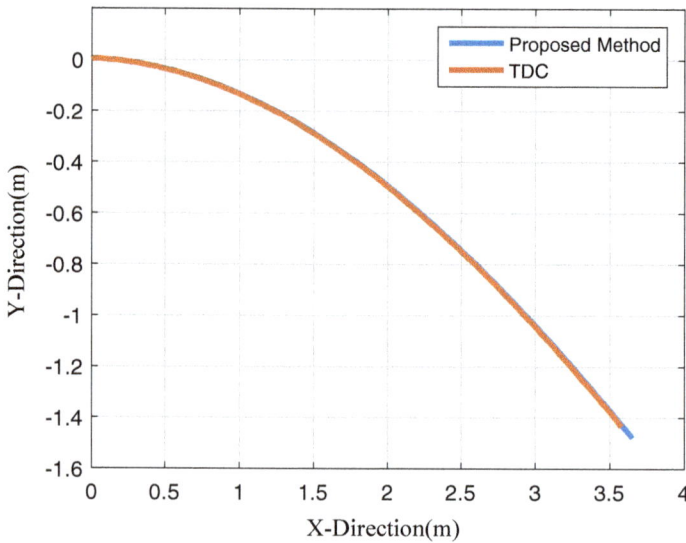

Fig. 4.2 Global trajectory

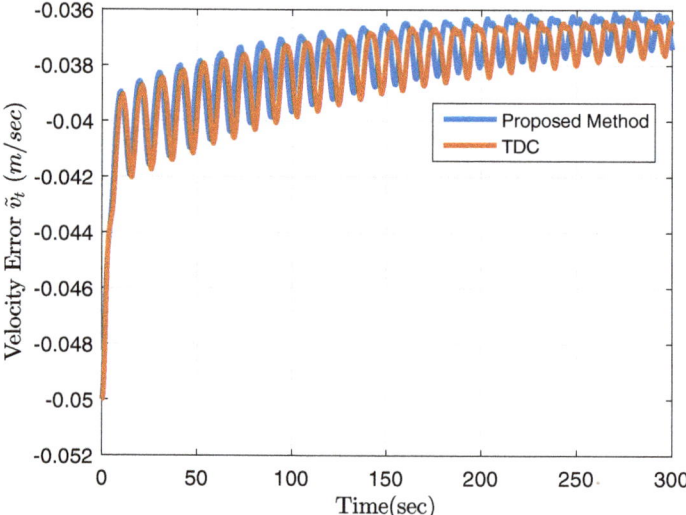

Fig. 4.3 Tangential velocity error

by the ARTDC method. The corresponding gait-angle frequency and offset leading to the obtained velocity and head-angle are exhibited in Figs. 4.5 and 4.6, respectively. The output function of the inner layer and the VHCs of the outer layer are presented in Figs. 4.7 and 4.8 for both TDC and the ARTDC approach. The output function as well as the VHCs exhibit convergence to a lower stability bound by virtue of the

4.3 Simulation Results

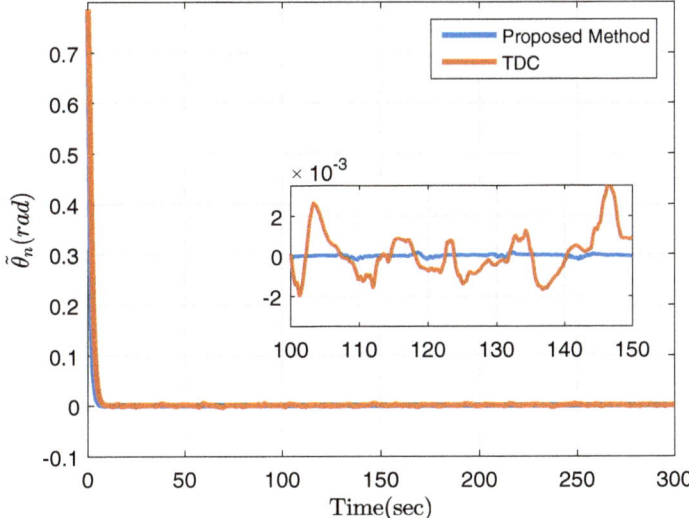

Fig. 4.4 Global head-angle error

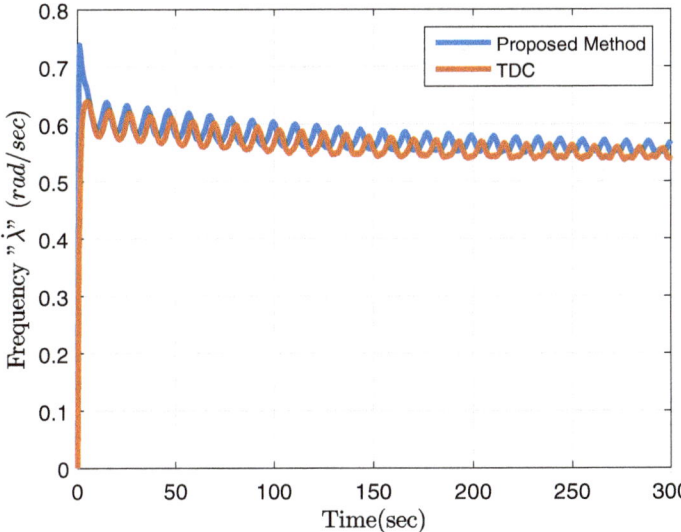

Fig. 4.5 Gait function frequency

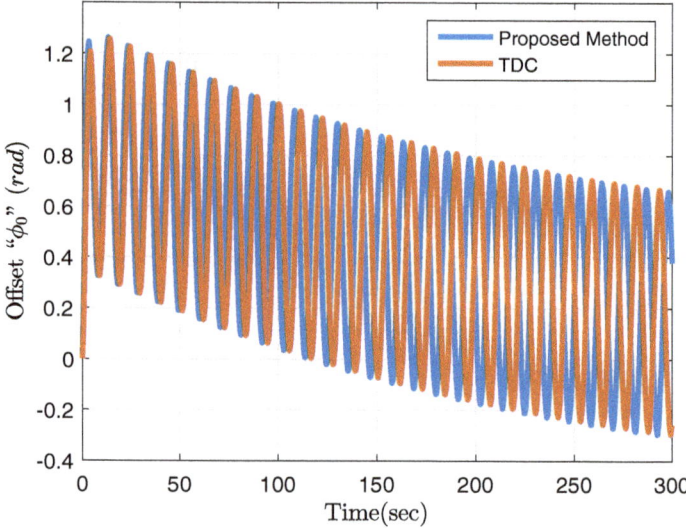

Fig. 4.6 Gait function offset

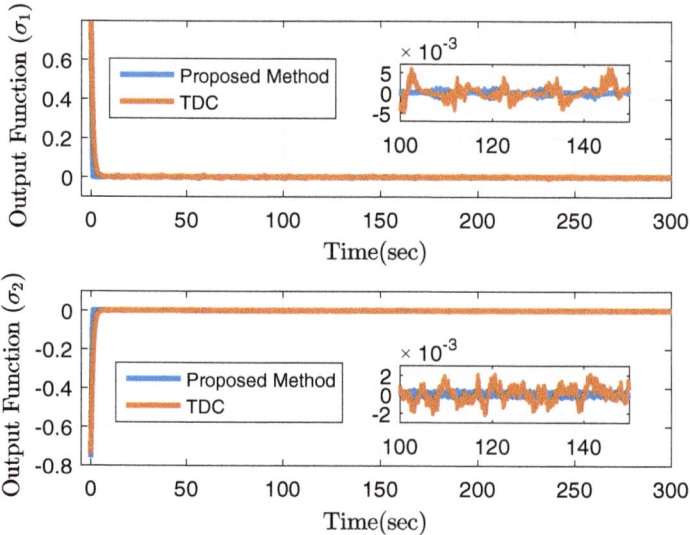

Fig. 4.7 Output function

4.3 Simulation Results

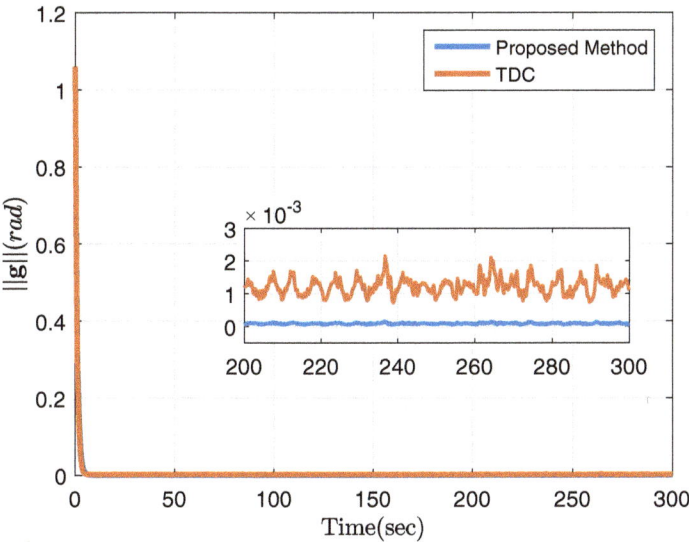

Fig. 4.8 Norm of VHCs

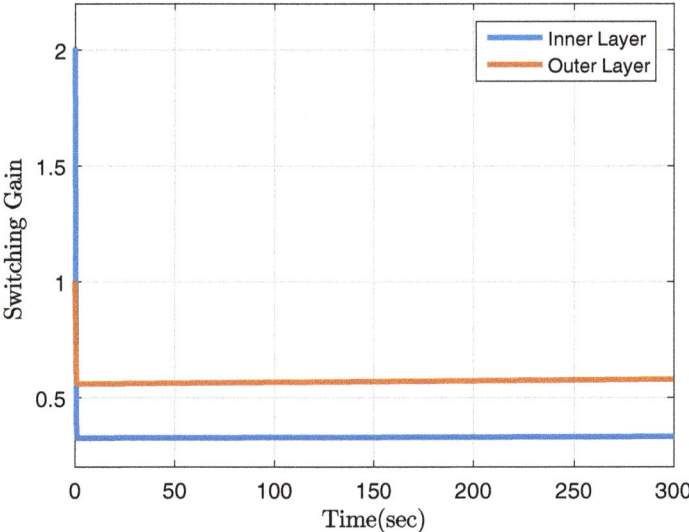

Fig. 4.9 Switching gain adaptation

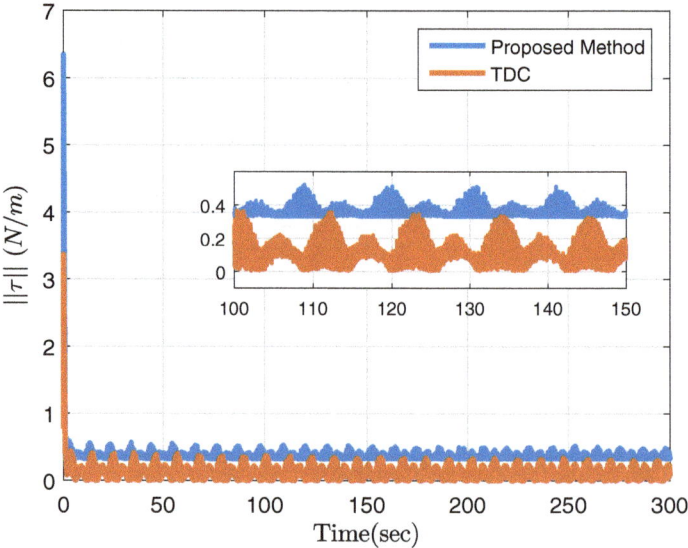

Fig. 4.10 Norm of control effort

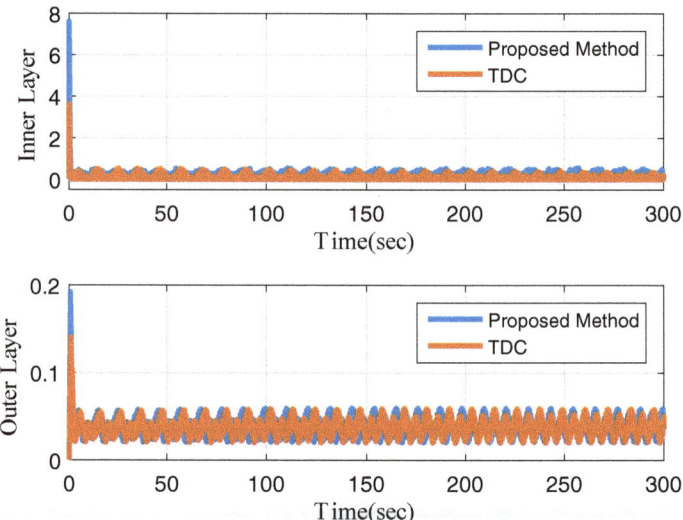

Fig. 4.11 Norm of estimation error

ARC. The variation of the switching gains for the inner and the outer layer is being shown in Fig. 4.9. The inner-layer switching gain η_σ and the outer-layer switching gain η_h initially decrease from the large initial conditions followed by fine adjustment to ensure improved tracking performance. The norm of the total control input has been presented in Fig. 4.10 which shows an initial surge which can be associated with the initial high switching gain. Also, the switching control law is responsible for the average control input to be higher for the proposed approach than the TDC-based controller. The estimation error from the inner- and outer-layer TDEs for both the ARTDC and TDC methodologies are of identical range as is shown in Fig. 4.11. The initial high value in the estimation error is due to the unavailability of input–output data for estimation. As the input–output data becomes available to the TDE, the error reduces to a lower bound thus exemplifying the accuracy of the TDE. It is to be noted, that the ARTDC methodology provides improved performance for similar estimation errors without any significant increment in the control input.

4.4 Summary

The work in this chapter proposes a double-layered ARTDC scheme for a planar snake robot. The outer-layer ARTDC handles the attainment of a particular body shape critical to generate required motion. Likewise, the inner-loop ARTDC attempts to achieve a predefined head-angle and velocity. The ARTDC approach mainly involves a TDE that estimates the uncertainties in the system from the immediate past input and output data. In addition to the feedback control law involving the uncertainty estimates, a switching control law has been induced to further stabilize the system in the presence of TDE error. The proposed approach circumvents the overestimation problem of switching gain without any prior knowledge of the uncertainty bound. Moreover, unlike the conventional adaptive sliding mode schemes, ARTDC does not presume the uncertainties to be bounded by a constant. Lyapunov stability analysis proves UUB stability of the closed-loop system and also presents guidelines for appropriate tuning of control gains. Simulation studies demonstrate the superior performance of the proposed approach in comparison to the TDC approach employed for the same system.

References

1. Mukherjee, J., Roy, S., Kar, I.N., Mukherjee, S.: A double-layered artificial delay-based approach for maneuvering control of planar snake robots. J. Dyn. Syst., Meas., Control **141**(4) (2018). https://doi.org/10.1115/1.4042033
2. Roy, S., Kar, I.N.: Adaptive-robust control of uncertain Euler-Lagrange systems with past data: a time-delayed approach. In: 2016 IEEE International Conference on Robotics and Automation (ICRA), pp. 5715–5720 (2016). https://doi.org/10.1109/ICRA.2016.7487795

3. Roy, S., Kar, I.N., Lee, J., Jin, M.: Adaptive-robust time-delay control for a class of uncertain Euler-Lagrange systems. IEEE Trans. Ind. Electron. **64**(9), 7109–7119 (2017)
4. Mukherjee, J., Roy, S., Kar, I.N., Mukherjee, S.: Maneuvering control of planar snake robot: An adaptive robust approach with artificial time delay. Int. J. Robust Nonlinear Control (2021). https://doi.org/10.1002/rnc.5430 [In Press]
5. Mukherjee, J., Mukherjee, S., Kar, I.N.: Sliding mode control of planar snake robot with uncertainty using virtual holonomic constraints. IEEE Robot. Autom. Lett. **2**(2), 1077–1084 (2017). https://doi.org/10.1109/LRA.2017.2657892
6. Mukherjee, J., Kar, I.N., Mukherjee, S.: Adaptive sliding mode control for head-angle and velocity tracking of planar snake robot. In: 2017 11th Asian Control Conference (ASCC), pp. 537–542 (2017). https://doi.org/10.1109/ASCC.2017.8287227

Chapter 5
Differential Flatness and Its Application to Snake Robots

Abstract Snake robot is a complicated dynamical system comprised of large number of inputs and generalized coordinates. Moreover, the dynamical equations of motion are highly nonlinear and coupled. The aforementioned complexities make path planning and control system design particularly difficult for this kind of system. The control approaches presented in Chaps. 1, 2, 3 and 4 confirm this observation as they employ a multi-layer control methodology to compute the torque input which is not directly mapped to the output space trajectory of the robot. Furthermore, dynamically capable path planning is still beyond the reach for snake robots. Hence, to address these issues and go beyond the conventional approach, it is required to map the control inputs directly to the output variables. This can be accomplished by adopting a differential-flatness-based approach. Differential flatness is a method of transforming a nonlinear system into a *flat system*, i.e. a corresponding system of flat outputs by establishing a diffeomorphic relation between the states and the outputs [1–3]. The flat outputs chosen should be either measurable or be computed from the measured variables. Therefore, trajectories designed as functions of flat outputs become convenient and the tracking control problem becomes simpler as well.

Mostly underactuated systems have been found to be flat but not all. Wheeled mobile robot is a very common flat system as shown in [4, 5] and have been widely researched to obtain effective control schemes to achieve tracking. Trailer systems [6] and induction motor [7] are also common examples of flat system. Recent works also cite that underactuated robotic manipulators can also be flat but their flatness property depend largely on their inertia distribution [8]. Most of the work in literature concern manipulators with single underactuation or one passive joint.

The work presented in this chapter proposes an approach to directly establish a relation between the output space variables of a snake robot to its control inputs. This will simplify the control problem as a control law can be designed with ease to track a desired output trajectory. An appropriate output variable is to be chosen which can capture the whole essence of robot motion. In the work presented in this chapter, flatness has been established for a snake robot using the global coordinates of the robot CM transformed to the head-link frame as flat output. These approach requires the basic assumption that all the generalized coordinate and their finite higher order derivatives are measurable or determinable. A flatness-based control approach has been proposed for wheeled mobile robot (WMR) with trajectory errors as output [9]. This alternate choice of flat output proves that there may exist multiple possible set of flat output for the same system. A flatness-based approach utilizing the serpenoid gait has been proposed to obtain a reduced-order flat system which has been further utilized to design a robust control law for the entire system.

The rest of the chapter is arranged as Sect. 5.1 introduces the concept of differential flatness; The flatness property of a wheeled mobile robot with different flat outputs has been discussed in Sect. 5.2; A differentially flat system for the planar snake robot has been obtained utilizing the serpenoid gait function in Sect. 5.3; Sect. 5.4 presents a feedback control law design employing the flat system for planar snake robot; The flat system has been further utilized to design an adaptive robust control law in Sect. 5.5 to address uncertainties as defined in Chap. 2; Finally, the chapter is summarized in Sect. 5.6.

5.1 Brief on Differential Flatness

Flatness is a property or characteristic of a particular system according to which all the solutions of the system can be parameterized by finite number of functions and their derivatives [1–3]. This mathematical property is being extensively used for analysis and design of controllers for nonlinear systems possessing this character. From a control perspective, a simpler explanation of differential flatness for any nonlinear system of the form

$$\dot{\mathbf{x}} = f(\mathbf{x}, \mathbf{u}), \qquad (5.1)$$

where $\mathbf{x} \in \mathbb{R}^n$ and $\mathbf{u} \in \mathbb{R}^m$ can be provided. The system can be stated to be differentially flat if there exists a set of parameters called *flat outputs* given as $\mathbf{y} \in \mathbb{R}^m$ such that

$$\begin{aligned}
\mathbf{x} &= f_1(\mathbf{y}, \dot{\mathbf{y}}, \ddot{\mathbf{y}}, ...), \\
\mathbf{u} &= f_2(\mathbf{y}, \dot{\mathbf{y}}, \ddot{\mathbf{y}}, ...), \\
\mathbf{y} &= f_3(\mathbf{x}, \mathbf{u}, \dot{\mathbf{u}}, \ddot{\mathbf{u}}, ...),
\end{aligned} \qquad (5.2)$$

i.e. the state and the input are functions of the flat outputs and its finite number of derivatives and the flat outputs are functions of state, input and finite number of derivatives of the input. The minimum number of times the flat outputs are needed to be differentiated to express the state and the input is called the *differential weight*. There may be different differential weights for different combination of flat outputs. During such cases, the minimum of those is considered as the actual differential weight.

5.1 Brief on Differential Flatness

Tasks like control design and path planning for a flat system, as the trajectory can be defined in terms of the flat outputs and the required control input can be obtained using the flatness property. Though flatness is closely related to feedback linearization, it would not be right to conceive flatness as method to achieve feedback linearization. It has been seen that flat systems with only a single input are static feedback linearizable, i.e. their differential weights are $n+1$. The differential weight of any static feedback linearizable system is $n+m$. For multi-input systems, the differential weight is more than that of $n+m$ and require prolongation to make them static feedback linearizable. Prolongation is a crucial method used for expressing a particular system as differentially flat. A flat system with differential weight more than $n+m$ have to be dynamically feedback linearizable.

5.2 Flatness for Wheeled Mobile Robots

This section presents the kinematic model of a differentially driven wheeled mobile robot (WMR) and its subsequent transformation into a body-fixed frame error model. Finally the flatness property of the error model has been established and the corresponding flat system has been derived.

5.2.1 Robot Kinematics

Figure 5.1 shows the schematic diagram of a two-wheeled differentially driven mobile robot, where (x, y) is the position of the robot's Center of Mass (CM) and θ is the orientation of the robot with respect to the global reference frame. For simplic-

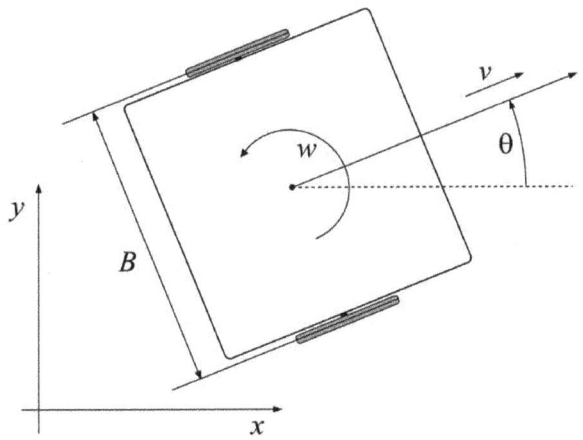

Fig. 5.1 Schematic diagram of the differentially driven WMR

ity, the wheel-axis-center has been considered to coincide with the robot CM. The kinematic equations map the derivative of the generalized coordinate vector to the linear and angular robot velocities. The kinematic equations can be expressed as

$$\dot{\mathbf{q}} = \begin{bmatrix} \dot{x} \\ \dot{y} \\ \dot{\theta} \end{bmatrix} = \begin{bmatrix} \cos(\theta) & 0 \\ \sin(\theta) & 0 \\ 0 & 1 \end{bmatrix} \begin{bmatrix} v \\ \omega \end{bmatrix}, \quad (5.3)$$

where $\mathbf{q} = [x \ y \ \theta]^T$ is called the generalized coordinate vector, v is the linear velocity and ω is the angular velocity. The velocities of right and left wheel can be obtained as

$$v_r = v + \frac{\omega B}{2} \ ; \ v_l = v - \frac{\omega B}{2},$$

where B is the inter-wheel distance. The nonholonomic constraint owing to no sideslip of the robot can be expressed as

$$\begin{bmatrix} -\sin\theta(t) & \cos\theta(t) \end{bmatrix} \begin{bmatrix} \dot{x}(t) \\ \dot{y}(t) \end{bmatrix} = 0. \quad (5.4)$$

5.2.2 Error Model of Robot Kinematics

In a trajectory tracking problem, the robot is supposed to track a desired reference trajectory. The reference trajectory is defined as

$$\mathbf{q_r}(t) = \begin{bmatrix} x_r(t) & y_r(t) & \omega_r(t) \end{bmatrix}^T. \quad (5.5)$$

The objective of the controller is to minimize the error between the desired and actual trajectory. Usually, the error modeling is done in the global reference frame. In a robot, the internal sensors provide data in the body-fixed frame and these measurements are then utilized to calculate the global error by some transformation. Thus, if a controller can be designed without having such transformations, it is supposed to be a much better approach provided the controller results in satisfactory performance. In the present work, the robot model have been expressed in terms of the posture-error utilizing the robot kinematics. The posture-error defined in the body-fixed frame is given as [10]

$$\begin{bmatrix} e_x \\ e_y \\ e_\theta \end{bmatrix} = \begin{bmatrix} \cos\theta & \sin\theta & 0 \\ -\sin\theta & \cos\theta & 0 \\ 0 & 0 & 1 \end{bmatrix} (\mathbf{q_r} - \mathbf{q}). \quad (5.6)$$

5.2 Flatness for Wheeled Mobile Robots

The kinematic posture-error model in body-fixed frame is obtained by combining (5.3) and (5.6) and is given as

$$\begin{bmatrix} \dot{e}_x \\ \dot{e}_y \\ \dot{e}_\theta \end{bmatrix} = \begin{bmatrix} \cos e_\theta & 0 \\ \sin e_\theta & 0 \\ 0 & 1 \end{bmatrix} \begin{bmatrix} v_r \\ \omega_r \end{bmatrix} + \begin{bmatrix} -1 & e_y \\ 0 & -e_x \\ 0 & -1 \end{bmatrix} \mathbf{u}, \tag{5.7}$$

where v_r and ω_r are the reference linear and angular velocity. This model can be used to design a suitable controller to minimize the errors and achieve tracking. The structure of control law for the trajectory tracking problem for such a model can be written [11] as

$$\mathbf{u} = \begin{bmatrix} v \\ \omega \end{bmatrix} = \begin{bmatrix} v_r \cos e_\theta + v_b \\ \omega_r + \omega_b \end{bmatrix}, \tag{5.8}$$

where $[v_b \ \omega_b]^T$ is the feedback control input to achieve tracking. Utilizing (5.8) in (5.7), the error model becomes

$$\begin{bmatrix} \dot{e}_x \\ \dot{e}_y \\ \dot{e}_\theta \end{bmatrix} = \begin{bmatrix} \omega_r e_y \\ -\omega_r e_x + v_r \sin e_\theta \\ 0 \end{bmatrix} + \begin{bmatrix} -1 & e_y \\ 0 & -e_x \\ 0 & -1 \end{bmatrix} \begin{bmatrix} v_b \\ \omega_b \end{bmatrix}. \tag{5.9}$$

5.2.3 Error Model as Flat System

One needs to first verify the flatness property of a particular system and find out the corresponding flat outputs as well. Though WMR is a very common example of flat system, it is extremely important to verify whether the presented error model is flat. The model given (5.9) can be expressed as

$$\dot{\mathbf{e}} = \mathbf{f}(\mathbf{e}) + \mathbf{g}(\mathbf{e})\mathbf{u_b}, \tag{5.10}$$

where $\mathbf{e} = [e_x \ e_y \ e_\theta]^T$ and $\mathbf{u_b} = [v_b \ \omega_b]^T$. In this approach, the longitudinal and lateral errors, (e_x, e_y), have been considered to be the flat outputs as

$$\mathbf{F} = \begin{bmatrix} F_1 \\ F_2 \end{bmatrix} = \begin{bmatrix} e_x \\ e_y \end{bmatrix}. \tag{5.11}$$

Consecutively, the error vector can be mapped to the flat outputs and their derivatives as [9]

$$e_x = \begin{bmatrix} 1 & 0 \end{bmatrix} \mathbf{F}, \quad e_y = \begin{bmatrix} 0 & 1 \end{bmatrix} \mathbf{F},$$
$$e_\theta = atan\left(\frac{\dot{F}_2/\dot{F}_1}{1 + p_r^2 - p_r(\dot{F}_2/\dot{F}_1)}\right), \quad (5.12)$$

where $p_r = (\dot{y}_r/\dot{x}_r)$.

Moreover, the inputs can be mapped as functions of the flat outputs and their derivatives as [9]

$$\omega_b = -\dot{e}_\theta = -\frac{(1 + p_r^2)(\dot{F}_1\ddot{F}_2 - \ddot{F}_1\dot{F}_2)}{((1 + p_r^2)\dot{F}_1 - p_r\dot{F}_2)^2 + \dot{F}_2^2},$$
$$v_b = \omega_r F_2 - \dot{F}_1 - F_2 \frac{(1 + p_r^2)(\dot{F}_1\ddot{F}_2 - \ddot{F}_1\dot{F}_2)}{((1 + p_r^2)\dot{F}_1 - p_r\dot{F}_2)^2 + \dot{F}_2^2}. \quad (5.13)$$

Using (5.11), (5.12) and (5.13), the error system (5.9) can be claimed to be differentially flat. As the mapping between the inputs and flat outputs is not singular, no form of prolongation is required for this system. The flat system corresponding to the posture-error model of the robot can be expressed as [9]

$$\dot{\mathbf{F}} = \begin{bmatrix} \dot{F}_1 \\ \dot{F}_2 \end{bmatrix} = \begin{bmatrix} \omega_r F_2 \\ -\omega_r F_1 + v_r sine_\theta \end{bmatrix} + \begin{bmatrix} -1 & F_2 \\ 0 & -F_1 \end{bmatrix} \begin{bmatrix} v_b \\ w_b \end{bmatrix}. \quad (5.14)$$

Remark: This part of the work is pivotal in explaining the flatness property of wheeled mobile robot with different combination of flat output. As most of the practical robots take velocity as input, it becomes extremely useful to work with the kinematic model using which a linear and angular velocity input can be designed. Another important aspect of flatness used in mobile robots is that path planning can be easily done in terms of the flat outputs. We expect that this flat system would be extremely useful to work with in cases where the error needs to converge in a desired manner. In addition, it is also notable that this model does not require any prolongation where the dimension of the error vector remains same. In a way we can say that this system is complimentary to that of the flat system with global position (x, y) as flat output.

In the subsequent sections, we have evaluated the characteristics of such a choice of flat output and the corresponding flat system. With the flat system derived, it is possible to design a very simple controller to solve the trajectory tracking problem.

5.2.4 Flatness-Based Control Law

In this section, we present a simple control law that results in efficient trajectory tracking. The linear and angular feedback control input has been chosen as

5.2 Flatness for Wheeled Mobile Robots

$$\omega_b = -\omega_r + \left(v_r \frac{\sin e_\theta}{F_1}\right) + a_3 + a_4\left(\frac{F_2}{F_1}\right),$$
$$v_b = \omega_r F_2 + F_2 \omega_b + a_1 F_1 + a_2 F_2, \tag{5.15}$$

where a_1, a_2, a_3 and a_4 are scalar constants. Applying the control law (5.15) to the flat system (5.14), the closed-loop system can be obtained as

$$\begin{bmatrix} \dot{e}_x \\ \dot{e}_y \end{bmatrix} = -\begin{bmatrix} a_1 & a_2 \\ a_3 & a_4 \end{bmatrix}\begin{bmatrix} e_x \\ e_y \end{bmatrix} = \mathbf{A}_c \begin{bmatrix} e_x \\ e_y \end{bmatrix},$$
$$\Rightarrow \dot{\mathbf{F}} = \mathbf{A}_c \mathbf{F}. \tag{5.16}$$

One needs to choose the gains a_1, a_2, a_3 and a_4 such that the matrix $\mathbf{A}_c$ becomes **Hurwitz**.

Remark: The control objective is to achieve, $(e_x, e_y) \to 0$ at $t \to \infty$. But $e_x = F_1 = 0$ is a *Singular Point* which limits F_1 from reaching to zero, i.e. F_1 is bounded as

$$F_1 \in (-\infty, -\epsilon] \cup [\epsilon, \infty), \tag{5.17}$$

where ϵ is a small positive constant.

As the flat output F_1 approach the region close to the equilibrium, i.e. $(-\epsilon, \epsilon)$, the control input is kept equal to the negative of reference input. In this case, the closed-loop equation becomes

$$\dot{\mathbf{F}} = \begin{bmatrix} v_r \\ v_r \sin e_\theta \end{bmatrix}. \tag{5.18}$$

The subsequent sections include the performance analysis of the proposed approach Lyapunov's method and the simulation studies as well.

5.2.5 Stability Analysis

A simple choice of Lyapunov function is taken as

$$V = \mathbf{F}^T \mathbf{P} \mathbf{F}, \tag{5.19}$$

where $\mathbf{P} = \mathbf{P}^T > 0$ is the solution of the Lyapunov equation given as

$$\mathbf{A}_c^T \mathbf{P} + \mathbf{P} \mathbf{A}_c = -\mathbf{Q}, \tag{5.20}$$

where $\mathbf{Q}$ is a positive definite symmetric matrix, i.e. $\mathbf{Q} = \mathbf{Q}^T > 0$. Differentiating the Lyapunov function (5.19) with respect to time yields

$$\dot{V} = \mathbf{F}^T \mathbf{P} \dot{\mathbf{F}} + \dot{\mathbf{F}}^T \mathbf{P} \mathbf{F}. \qquad (5.21)$$

The subsequent analysis can be categorized into two cases depending upon the control law based upon the value of F_1.

Case I: $|F_1| \geq \epsilon$
Utilizing closed-loop system (5.16) and the Lyapunov equation (5.20) in (5.21), $\dot{V}$ becomes

$$\dot{V} = \mathbf{F}^T (\mathbf{A_c}^T \mathbf{P} + \mathbf{P} \mathbf{A_c}) \mathbf{F} = -\mathbf{F}^T \mathbf{Q} \mathbf{F}.$$

As $\mathbf{Q}$ is a positive definite matrix

$$\dot{V} = -\mathbf{F}^T \mathbf{Q} \mathbf{F} < \mathbf{0}. \qquad (5.22)$$

As V is a monotonically decreasing function within this range, the state trajectory converges below the boundary value and doesn't increase beyond the boundary thereafter.

Case II: $|F_1| < \epsilon$
Utilizing the closed-loop system (5.18) in (5.21), one can obtain

$$\dot{V} = v_r (F_1 + F_2 \sin e_\theta).$$

The tracking convergence in this case cannot be confirmed. As a result, asymptotic convergence for the entire trajectory cannot be ensured as V is no more a monotonically decreasing function within the bound $(-\epsilon, \epsilon)$. But the error is bound to remain in the vicinity of the equilibrium point within this bound. Thus, the proposed control scheme results in an Uniformly Ultimately Bounded (UUB) stability where the error doesn't converge to origin, rather remains within an acceptable upper bound.

5.2.6 Simulation Results

Simulations were performed to analyze the efficiency of the control scheme proposed. Being a kinematic control, the robot parameters doesn't affect the control law which is simpler to implement. The simulations were performed for a Circular and a Cubic-Spline Trajectory. The circular trajectory has been considered as

$$x_r = \sin(0.4t),$$
$$y_r = \cos(0.4t),$$

5.2 Flatness for Wheeled Mobile Robots

where t is time. The cubic-spline reference trajectory has been defined as

$$x_r = 0.15t - 0.01125t^2 + 0.000375t^3,$$
$$y_r = 0.009t^2 - 0.0003t^3.$$

Initial error and the bound on ϵ for both the simulation cases has been taken as

$$\begin{bmatrix} F_1(0) \\ F_2(0) \end{bmatrix} = \begin{bmatrix} 0.1 \\ 0.2 \end{bmatrix},$$
$$\epsilon = 5 \times 10^{-4}.$$

The simulation time for both the trajectories is 20 s. The scalar constants of the controller are chosen as [9],

$$\mathbf{A_c} = -\begin{bmatrix} a_1 & a_2 \\ a_3 & a_4 \end{bmatrix} = -\begin{bmatrix} 2 & 3 \\ -2 & 7 \end{bmatrix}.$$

Figures 5.2 and 5.5 shows the reference and actual trajectories of the circular and cubic-spline trajectories in global frame. These plots show that the actual trajectory convergence to the actual global trajectory even with non-zero initial error. The longitudinal and lateral error are the flat outputs and the corresponding flat system is expressed in (5.16). Figures 5.3 and 5.6 shows the flat errors for the two trajectories. The position error in the global frame for the two trajectories have been computed

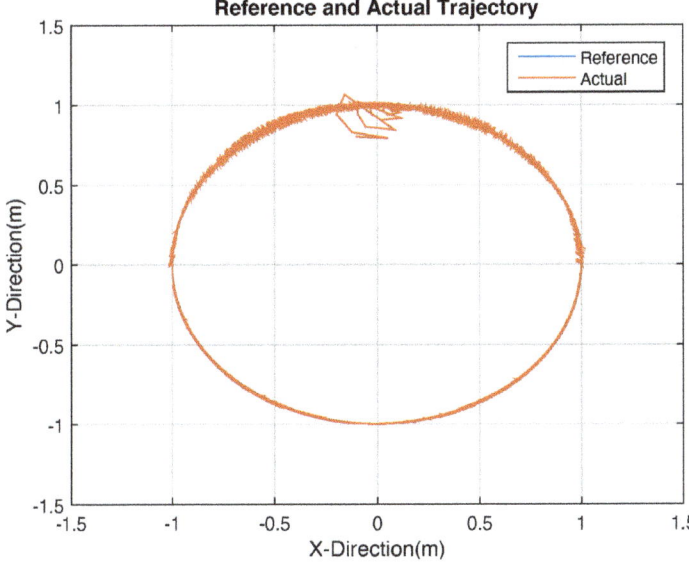

Fig. 5.2 Reference and actual trajectory of the robot

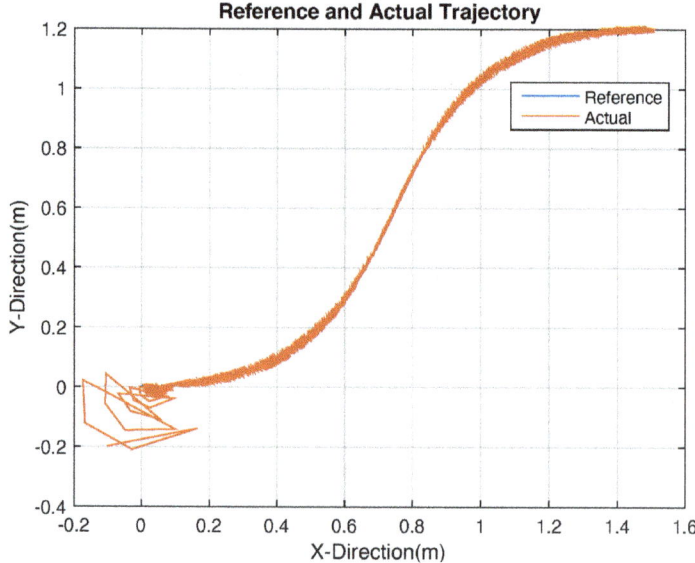

Fig. 5.3 Flat outputs

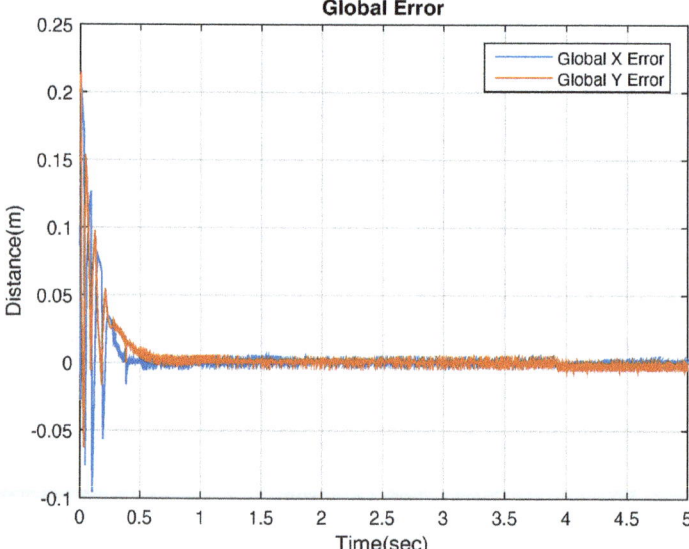

Fig. 5.4 Global frame error of the robot

5.2 Flatness for Wheeled Mobile Robots

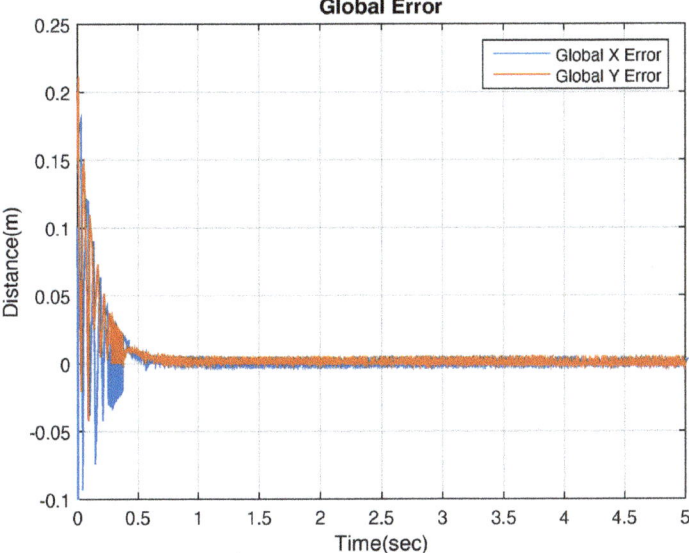

Fig. 5.5 Reference and actual trajectory of the robot

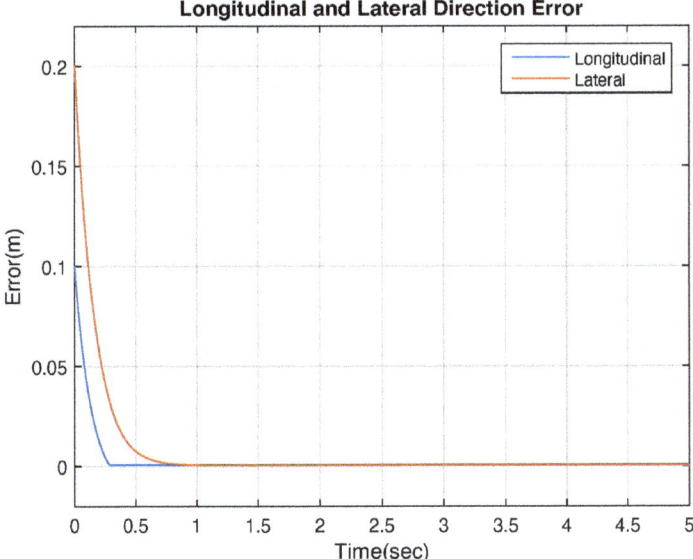

Fig. 5.6 Flat outputs

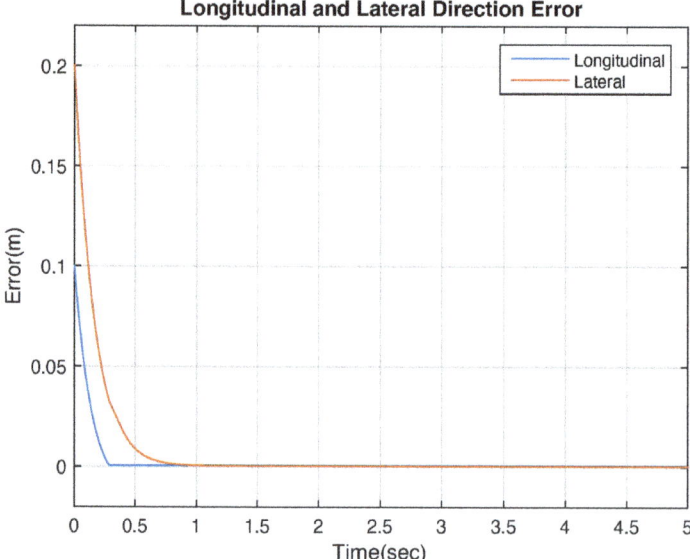

Fig. 5.7 Global frame error of the robot of the robot

using (5.6) which is shown in Figs. 5.4 and 5.7. The simulation results clearly show that the flat outputs as well as the global positional errors converge to a bound and remain with it. These results confirm theoretically proven UUB stability of the system.

5.3 Flatness of Snake Robot Utilizing Serpenoid Gait

Analyzing flatness by establishing mapping between the generalized coordinates for a complex dynamical system like snake robots is a tedious task. In literature, this has been implemented through the imposition of specific gait function, which also enforces a specific body shape. In this section, the serpenoid gait function already discussed in Sect. 1.5.1 has been utilized to obtain differential flatness for planar snake robots.

5.3.1 Establishing Flatness

The methodologies presented in the previous chapters have utilized a VHC-based approach to track the relative joint angles to the serpenoid gait function. Now, the error in the relative joint angle have been defined to be the outputs $\mathbf{F}_t$ expressed as

5.3 Flatness of Snake Robot Utilizing Serpenoid Gait

$$\mathbf{F}_t = \begin{bmatrix} F_1 & F_2 & \ldots & F_{n-1} \end{bmatrix}^T = \mathbf{D}\boldsymbol{\theta} - \boldsymbol{\Phi}(\lambda) - \mathbf{b}_1\phi_0. \quad (5.23)$$

To establish flatness, one needs to demonstrate a mapping to recreate the system states from the flat outputs and their higher order derivatives. In case of a snake robot, the generalized coordinates can be expressed as a function of the outputs as

$$\boldsymbol{\theta} = \mathbf{e}\theta_n + \mathbf{G}\left\{\mathbf{F}_t + \underbrace{\boldsymbol{\Phi}(\lambda) + \mathbf{b}_1\phi_0}_{\text{Gait function}}\right\}. \quad (5.24)$$

The angle of the head link in global frame θ_n also has to be obtained from the flat outputs, which will be discussed later. The torque inputs $\boldsymbol{\tau}$ driving the joints can also be directly mapped to the actuated output vector $\mathbf{F}_f$ as

$$\boldsymbol{\tau} = (\mathbf{D}\mathbf{M}^{-1}\mathbf{D}^T)^{-1}\left\{-\mathbf{D}\mathbf{M}^{-1}\left(\mathbf{W}\dot{\boldsymbol{\theta}}^2 + l\mathbf{S}\mathbf{C}_\theta^T\mathbf{f}_R\right) + \boldsymbol{\Phi}''(\lambda)\dot{\lambda}^2 + \boldsymbol{\Phi}'(\lambda)\ddot{\lambda} + \mathbf{b}_1\ddot{\phi}_0 - \ddot{\mathbf{F}}_t\right\}. \quad (5.25)$$

The matrices $\mathbf{M}$ and $\mathbf{W}$ are functions of vector $\boldsymbol{\theta}$ which can be obtained from the flat outputs using (5.24). However, the friction force vector is a function of the robot CM velocity $\dot{\mathbf{p}}$ in the global frame. These velocities are transformed to the body-fixed frame of the head link yielding the tangential and normal direction velocity of the robot CM as

$$\begin{bmatrix} v_t \\ v_n \end{bmatrix} = \begin{bmatrix} \cos\theta_n & \sin\theta_n \\ -\sin\theta_n & \cos\theta_n \end{bmatrix}\dot{\mathbf{p}}. \quad (5.26)$$

It is to be noted here that the velocity transformation to the head-link body frame ensures that the normal velocity v_n is always bounded owing to the anisotropic friction model described in (1.6). Therefore, the normal velocity is not required to be controlled, as it exhibits a limit cycle behavior about zero. The control objective remains to be the tracking of a specific heading θ_{ref} and velocity v_{ref}. Hence, the heading and velocity errors are defined as

$$\tilde{\theta}_n = \theta_n - \theta_{ref},$$
$$\tilde{v}_t = v_t - v_{ref}.$$

With the gait function considered in (1.28), it is known that the velocity can be controlled through gait frequency $\dot{\lambda}$, whereas the gait offset ϕ_0 is responsible for varying the heading [12]. Considering, the heading angle error as another flat output

$$F_n = \tilde{\theta}_n, \quad (5.27)$$

the system state of head angle can be obtained from the flat output as

$$\theta_n = F_n + \theta_{ref}. \tag{5.28}$$

With θ_n being a function of a flat output, the flatness condition (5.24) gets satisfied. It has already been discussed that choosing specific values of gait frequency and offset to achieve a specified velocity and heading is not straightforward. For this reason, a set of compensators are employed to obtain their values written as

$$\ddot{\lambda} = u_\lambda, \quad \ddot{\phi}_0 = u_{\phi_0},$$

where u_λ and u_{ϕ_0} are pseudo-inputs chosen to track the desired velocity and heading, respectively. However, the pseudo-input u_λ can not be directly mapped to the tangential velocity error and hence another output is being considered as

$$F_{n+1} = K_v \tilde{v}_t + \dot{\lambda}. \tag{5.29}$$

The tangential velocity can be computed from the flat output as

$$v_t = \frac{(F_{n+1} - \dot{\lambda})}{K_v} + v_{ref}. \tag{5.30}$$

Although, the normal velocity v_n is bounded and need not be controlled, it is measurable and can be considered as a pseudo-flat output without any demands on the control effort. Hence, with $\mathbf{F}_a = [F_n \ F_{n=1}]^T$ being the augmented flat outputs, robot CM velocity can be formulated as

$$\dot{\mathbf{p}} = \begin{bmatrix} \cos F_n & \sin F_n \\ -\sin F_n & \cos F_n \end{bmatrix} \begin{bmatrix} F_{n+1} \\ v_n \end{bmatrix}. \tag{5.31}$$

The pseudo-input u_λ can be expressed utilizing (1.36) as

$$\begin{aligned} u_\lambda &= \dot{F}_{n+1} - K_v \dot{\tilde{v}}_t \\ u_\lambda &= \dot{F}_{n+1} - K_v f_2(v_t, v_n, \theta_n, \dot{\theta}_n, \lambda, \dot{\lambda}, \phi_0, \dot{\phi}_0). \end{aligned} \tag{5.32}$$

Utilizing (5.28) and (5.30) in the previous equation, the pseudo-input for velocity tracking can be written in terms of the flat outputs as

$$u_\lambda = \dot{F}_{n+1} - K_v f_2(F_n, \dot{F}_n, F_{n+1}, v_n, \lambda, \dot{\lambda}, \phi_0, \dot{\phi}_0). \tag{5.33}$$

Furthermore, the pseudo-input u_{ϕ_0} can also be expressed using (1.38a) as

$$u_{\phi_0} = \frac{\ddot{F}_n + \ddot{\theta}_{\text{ref}} - \psi_1(\boldsymbol{\theta}, \dot{\boldsymbol{\theta}}, \dot{\mathbf{p}}, \lambda, \dot{\lambda}) - \psi_2(\boldsymbol{\theta}, \lambda) u_\lambda}{\psi_3(\boldsymbol{\theta})}. \tag{5.34}$$

5.3 Flatness of Snake Robot Utilizing Serpenoid Gait

Employing (5.28) and (5.30), the pseudo-flat output for heading can be mapped to the flat outputs as

$$u_{\phi_0} = \frac{\ddot{F}_n + \ddot{\theta}_{\text{ref}} - \psi_1(\mathbf{F}_t, \dot{\mathbf{F}}_t, \mathbf{F}_a, \dot{\mathbf{F}}_a, \lambda, \dot{\lambda}) - \psi_2(\mathbf{F}_t, \mathbf{F}_a, \lambda) u_\lambda}{\psi_3(\mathbf{F}_a, \mathbf{F}_u)}. \quad (5.35)$$

With the conditions for flatness being satisfied, one needs to obtain the dynamic equations for the flat output which can be employed for control design.

5.3.2 Flat System

The flat system with $\mathbf{F} = [\mathbf{F}_t^T \ \mathbf{F}_a^T]^T \in \mathbb{R}^{n+1}$ as flat outputs and $\mathbf{u} = [\boldsymbol{\tau}^T \ u_\lambda \ u_{\phi_0}]^T \in \mathbb{R}^{n+1}$ as the inputs can be expressed as

$$\begin{bmatrix} \ddot{\mathbf{F}}_t \\ \ddot{F}_n \\ \dot{F}_{n+1} \end{bmatrix} = \begin{bmatrix} \mathbf{DM}^{-1}\left(\mathbf{W}\dot{\theta}^2 + l\mathbf{SC}_\theta^T \mathbf{f}_R\right) - \boldsymbol{\Phi}''\dot{\lambda}^2 \\ \psi_1 - \ddot{\theta}_{\text{ref}} \\ K_v(f_2 - \dot{v}_{\text{ref}}) \end{bmatrix} + \begin{bmatrix} (\mathbf{DM}^{-1}\mathbf{D}^T) & -\mathbf{b} & -\boldsymbol{\Phi}' \\ \mathbf{0}_{1\times(n-1)} & \psi_3 & \psi_2 \\ \mathbf{0}_{1\times(n-1)} & 0 & 1 \end{bmatrix} \begin{bmatrix} \boldsymbol{\tau} \\ u_{\phi_0} \\ u_\lambda \end{bmatrix}. \quad (5.36)$$

It is to be noted here that the choice of the augmented flat outputs $\mathbf{F}_a$ depends upon the choice of the gait function. With chance of alternate gaits to generate snake motion, the choice of these flat outputs might change. Generally, the constraints in an underactuated system results in mathematical relations between various system states. However, for an overactuated system like the snake robot, such constraints are imposed on the body shape through the serpenoid gait.

With the flat system obtained, one can design various feedback laws to achieve desired motion for the snake robot. Moreover, uncertainties can be accommodated in the flat system through the design of robust as well as adaptive control laws. In this section, a feedback linearizing control law has been designed to demonstrate the control design for the derived flat system. A schematic block diagram of the flatness-based feedback control scheme is exhibited in Fig. 5.8.

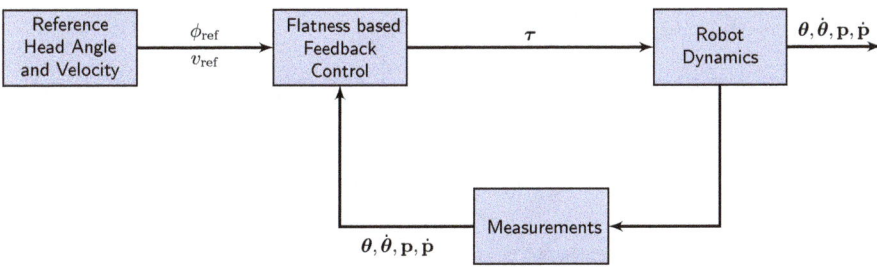

Fig. 5.8 Block diagram of flatness-based feedback control law

5.4 Feedback Control Law

With such a formulation, the tracking problem for the snake robot has been transformed in to a regulation problem, where the control objective is to regulate the flat outputs to zero. The flat system (5.36) can be written in state-space form as

$$\dot{\bar{\mathbf{F}}} = \bar{\mathbf{f}}_F + \bar{\mathbf{g}}_F \mathbf{u}_F, \tag{5.37}$$

where $\bar{\mathbf{F}} = \begin{bmatrix} \mathbf{F}_t^T & F_n & \dot{\mathbf{F}}_t^T & \dot{F}_n & F_{n+1} \end{bmatrix}^T \in \mathbb{R}^{2n+1}$ represents the state vector for the concatenated system, $\bar{\mathbf{f}}_F = \begin{bmatrix} \dot{\mathbf{F}}_t^T & \dot{F}_n & \mathbf{f}_F^T \end{bmatrix}^T$ and $\bar{\mathbf{g}}_F = \begin{bmatrix} \mathbf{0}_{n\times(n+1)}^T & \mathbf{g}_F^T \end{bmatrix}^T$. Notably, $\bar{\mathbf{g}}_F \in \mathbb{R}^{(2n+1)\times(n+1)}$ is the input matrix and $\mathbf{u}_F \in \mathbb{R}^{(n+1)\times 1}$ is the modified input matrix. Considering a completely known system, a simple feedback control law for the system (5.37) can be designed as

$$\mathbf{u}_F = (\bar{\mathbf{g}}_F^T \bar{\mathbf{g}}_F)^{-1} \left\{ -\bar{\mathbf{g}}_F^T \bar{\mathbf{f}}_F - \bar{\mathbf{g}}_F^T \mathbf{K}_F \bar{\mathbf{F}} \right\}, \tag{5.38}$$

where $\mathbf{K}_F \in \mathbb{R}^{(2n+1)\times(2n+1)}$ is a suitable positive definite gain matrix to be chosen. For the control law (5.38) to exist, it is essential that $(\mathbf{g}_F^T \mathbf{g}_F)$ must be invertible, i.e.

$$\det(\bar{\mathbf{g}}_F^T \bar{\mathbf{g}}_F) = \left| \begin{bmatrix} (\mathbf{DM}^{-1}\mathbf{D}^T)^T(\mathbf{DM}^{-1}\mathbf{D}^T) & (\mathbf{DM}^{-1}\mathbf{D}^T)^T \mathbf{b}_1 & (\mathbf{DM}^{-1}\mathbf{D}^T)^T \mathbf{\Phi}' \\ \mathbf{b}_1^T(\mathbf{DM}^{-1}\mathbf{D}^T) & \mathbf{b}_1^T \mathbf{b}_1 + \psi_3^2 & \mathbf{b}_1^T \mathbf{\Phi}' + \psi_2 \psi_3 \\ \mathbf{\Phi}'^T(\mathbf{DM}^{-1}\mathbf{D}^T) & \mathbf{\Phi}'^T \mathbf{b}_1 + \psi_2 \psi_3 & \mathbf{\Phi}'^T \mathbf{\Phi}' + \psi_2^2 \end{bmatrix} \right| \neq 0. \tag{5.39}$$

A particular structure for the gain matrix has been chosen as

$$\mathbf{K}_F = \begin{bmatrix} \mathbf{0}_{n\times n} & \mathbf{0}_{n\times(n+1)} \\ \hline \mathbf{K}_{F_2} & \mathbf{K}_{F_1} \end{bmatrix},$$

where

$$\mathbf{K}_{F_1} = \begin{bmatrix} K_{a_1}\mathbf{I}_{n-1} & \mathbf{0}_{(n-1)\times 1} & \mathbf{0}_{(n-1)\times 1} \\ \hline \mathbf{0}_{1\times(n-1)} & K_{n_1} & 0 \\ \hline \mathbf{0}_{1\times(n-1)} & 0 & K_{n+1} \end{bmatrix} \quad \text{and} \quad \mathbf{K}_{F_2} = \begin{bmatrix} K_{a_2}\mathbf{I}_{n-1} & \mathbf{0}_{(n-1)\times 1} \\ \hline \mathbf{0}_{1\times(n-1)} & K_{n_2} \\ \hline \mathbf{0}_{1\times(n-1)} & 0 \end{bmatrix}.$$

Utilizing the control law (5.38) in (5.37), the closed-loop flat system can be expressed as

$$\bar{\mathbf{g}}_F \left\{ \dot{\bar{\mathbf{F}}} + \mathbf{K}_F \bar{\mathbf{F}} \right\} = 0. \tag{5.40}$$

With further simplifications, the closed-loop system can be expressed as

5.4 Feedback Control Law

$$\ddot{\mathbf{F}}_t + K_{a_1}\dot{\mathbf{F}}_t + K_{a_2}\mathbf{F}_t = \mathbf{0}, \tag{5.41a}$$

$$\ddot{F}_n + K_{n_1}\dot{F}_n + K_{n_2}F_n = 0, \tag{5.41b}$$

$$\dot{F}_{n+1} + K_{n+1}F_{n+1} = 0. \tag{5.41c}$$

The following subsection details the stability analysis of the flat system through the control approach presented in this subsection.

5.4.1 Stability Analysis

Considering a Lyapunov function of the form

$$V_F = \frac{1}{2}\bar{\mathbf{F}}^T\bar{\mathbf{F}}. \tag{5.42}$$

Taking time derivative of the Lyapunov function (5.42) with respect to time, one obtains

$$\dot{V}_F = \bar{\mathbf{F}}^T\dot{\bar{\mathbf{F}}}. \tag{5.43}$$

The closed-loop system (5.41) can be expressed in the state-space form as

$$\dot{\bar{\mathbf{F}}} = \mathbf{A}_F\bar{\mathbf{F}}, \tag{5.44}$$

where

$$\mathbf{A}_F = \left[\begin{array}{c|c|c|c|c} \mathbf{0}_{(n-1)\times(n-1)} & \mathbf{0}_{(n-1)\times 1} & \mathbf{I}_{n-1} & \mathbf{0}_{(n-1)\times 1} & \mathbf{0}_{(n-1)\times 1} \\ \hline \mathbf{0}_{1\times(n-1)} & 0 & \mathbf{0}_{1\times(n-1)} & 1 & 0 \\ \hline -K_{a_2}\mathbf{I}_{n-1} & \mathbf{0}_{(n-1)\times 1} & -K_{a_1}\mathbf{I}_{n-1} & \mathbf{0}_{(n-1)\times 1} & \mathbf{0}_{(n-1)\times 1} \\ \hline \mathbf{0}_{1\times(n-1)} & -K_{n_2} & \mathbf{0}_{1\times(n-1)} & -K_{n_1} & 0 \\ \hline \mathbf{0}_{1\times(n-1)} & 0 & \mathbf{0}_{1\times(n-1)} & 0 & -K_{n+1} \end{array}\right].$$

The closed-loop system (5.44) has been further used in (5.43) to obtain

$$\dot{V}_F = \bar{\mathbf{F}}^T\mathbf{A}_F\bar{\mathbf{F}}. \tag{5.45}$$

The gains can be appropriately chosen so that $\mathbf{A}_F$ becomes a negative definite matrix with desired eigenvalues which results into

$$\dot{V}_F < 0. \tag{5.46}$$

This result proves exponential stability of the flat system.

5.4.2 Simulation Results

The performance of the planar snake robot through the proposed approach has been verified through simulation studies and compared with the state-of-the-art methodology [12]. The physical specifications of the snake robot considered are given in Table 2.1. The reference velocity and head angle along with the nominal friction coefficients are given in Table 5.1.

The system has been assumed to be completely known and no uncertainty or external disturbance have been considered. The various controller gains for the proposed control law are given in Table 5.2, whereas gains for the state-of-the-art approach are provided in Table 2.3.

5.4.3 Discussion

The trajectory of the robot CM utilizing the proposed flatness-based approach in comparison to the state-of-the-art methodology presented in Chap. 1 has been shown in Fig. 5.9. The superior trajectory tracking performance of the proposed method can be clearly inferred from this figure. The tangential velocity error in Fig. 5.10 and head-angle error in Fig. 5.11 also attest this observation through improvements by the proposed approach. The norm of the torque input employed for tracking the trajectory is given in Fig. 5.12. The steady-state control input for the proposed method can be seen to be comparable to the existing approach. The corresponding frequency and offset of the gait angle resulting into such velocity and heading performance are

Table 5.1 Reference and nominal values

Parameter	Numerical value
θ_{ref}	$-\pi/4$ rad
v_{ref}	0.05 m/s^2
$\hat{c}_t$	0.5
$\hat{c}_n$	3
α	$30\pi/180$ rad
δ	$72\pi/180$ rad

Table 5.2 Control parameters

Parameter	Numerical value
K_{a_1}	10
K_{a_2}	10
K_{n_1}	10
K_{n_2}	10
K_{n+1}	30
K_v	14

5.4 Feedback Control Law

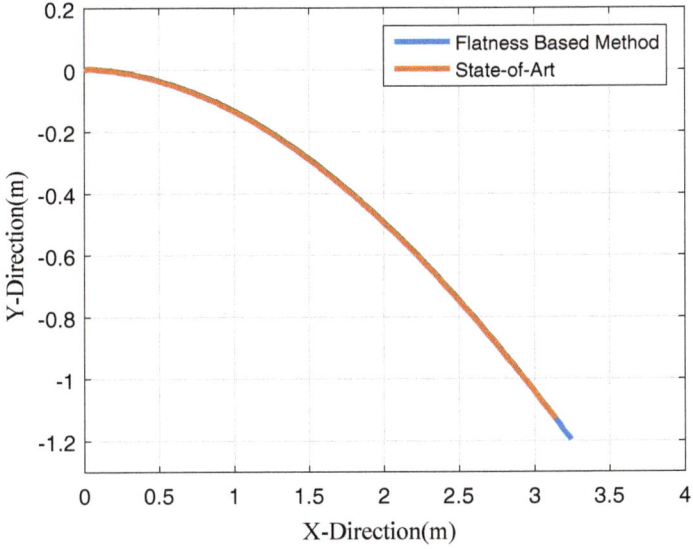

Fig. 5.9 Global trajectory

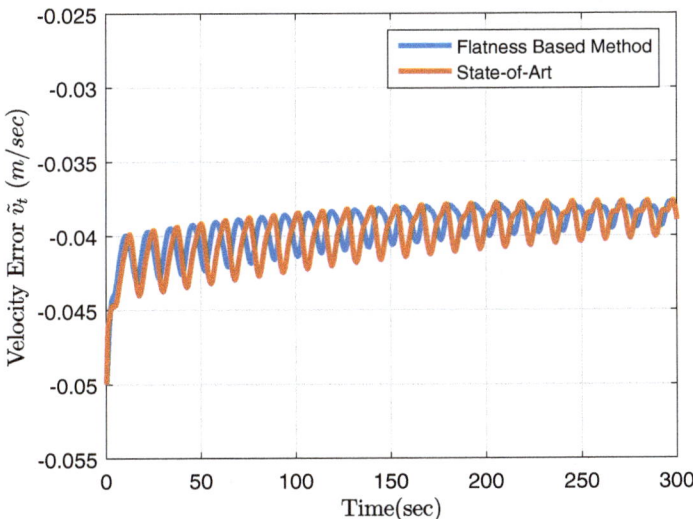

Fig. 5.10 Tangential velocity error

112　　　　　　　　　　　　　5　Differential Flatness and Its Application to Snake Robots

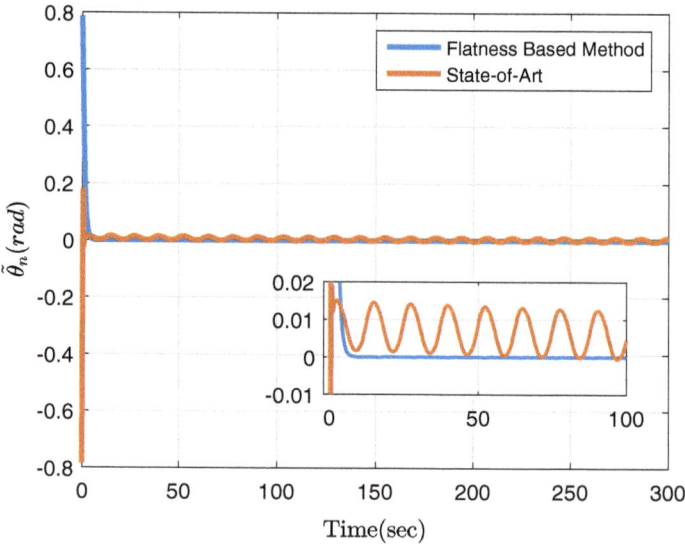

Fig. 5.11 Head-angle error

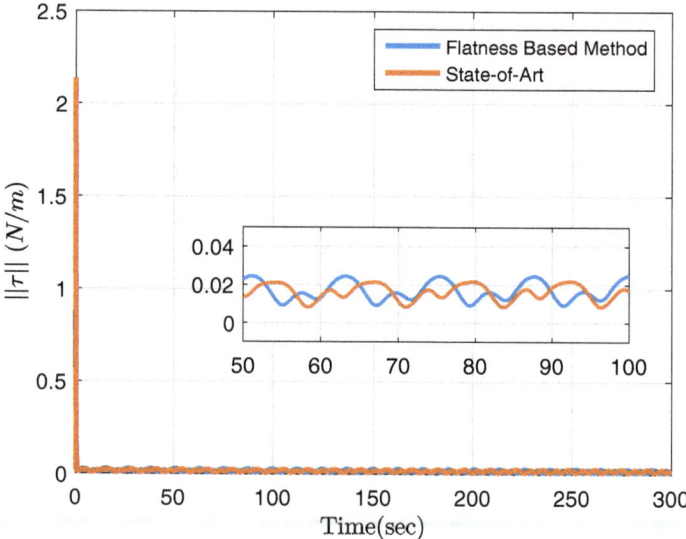

Fig. 5.12 Norm of control effort

5.4 Feedback Control Law

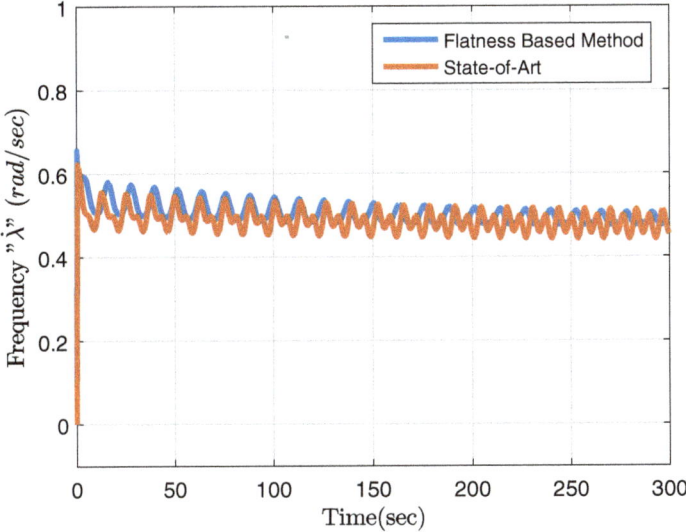

Fig. 5.13 Gait function frequency

shown in Fig. 5.13 and Fig. 5.14, respectively. The norm of the actuated as well the un-actuated flat outputs are presented in Fig. 5.15 and Fig. 5.16, respectively, which show bounded behavior for the flat outputs.

It is important to note that the effect of the actuated part of the flat output system on the heading and velocity subsystems has been considered in this work. Hence, the entire control law design can be executed in a single layer for the flat system which results in a desired performance of the whole snake robot.

5.5 Adaptive Robust Control Design for Flat Systems

The feedback control law presented in Sect. 5.4 doesn't perform satisfactorily for a system with unknown parameters. Hence, in this section an adaptive robust control law will be proposed for the flat system (5.35) considering parametric uncertainties in the system dynamics. A schematic block diagram of the flatness-based adaptive robust control scheme is exhibited in Fig. 5.17.

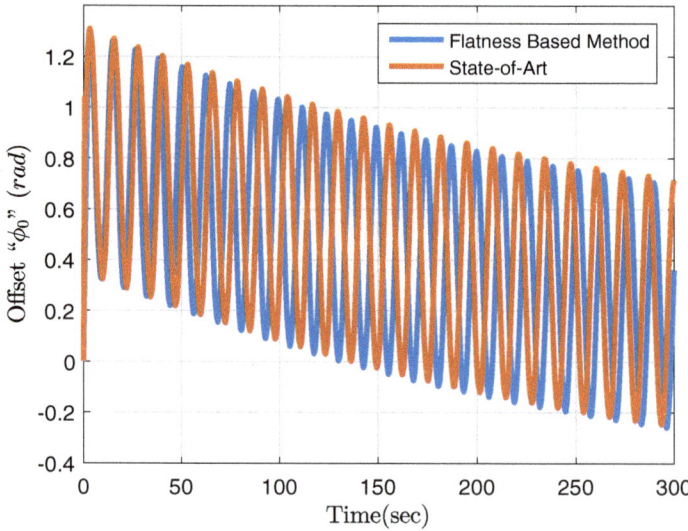

Fig. 5.14 Gait function offset

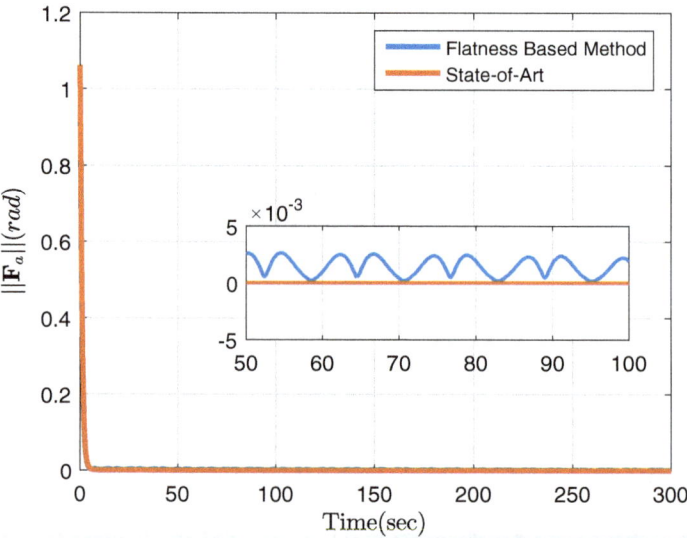

Fig. 5.15 Norm of actuated flat O/p

5.5 Adaptive Robust Control Design for Flat Systems

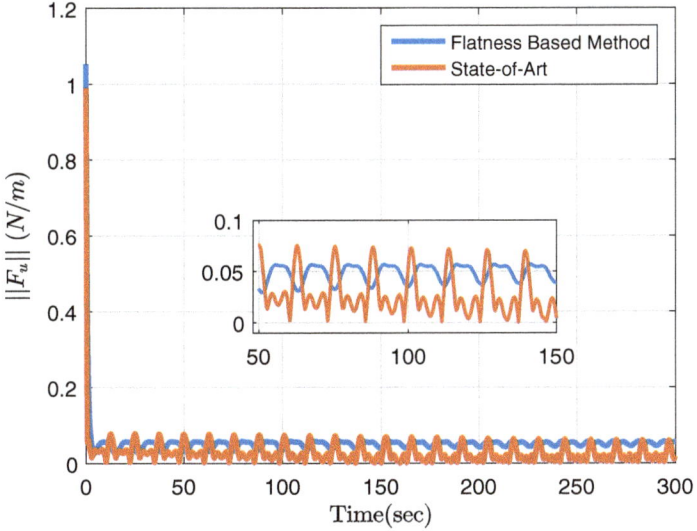

Fig. 5.16 Norm of augmented flat O/p

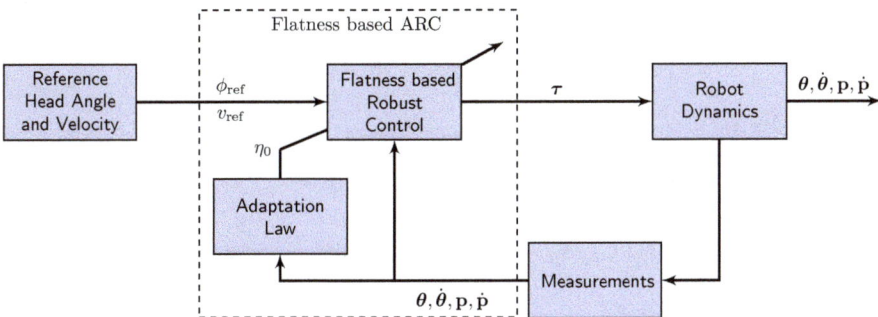

Fig. 5.17 Block diagram of flatness-based adaptive robust control law

5.5.1 Robust Control Law for Flat Systems with Uncertainties

Considering filter tracking error for the actuated flat outputs as

$$\bar{\mathbf{F}}_t = \dot{\mathbf{F}}_t + \mathbf{K}_t \mathbf{F}_t, \tag{5.47}$$

where $\mathbf{K}_t$ is a positive definite matrix to be appropriately chosen. Convergence of $\bar{\mathbf{F}}_t$ ensures the convergence of $\dot{\mathbf{F}}_t$ and $\mathbf{F}_t$ to the stable equilibrium point. Similarly, a filter tracking error is also considered for the head-angle error as

$$\bar{F}_n = \dot{F}_n + K_n F_n, \tag{5.48}$$

where K_n is chosen to be positive. Hence, the overall flat system with $\mathbf{F} = [\bar{\mathbf{F}}_t^T \ \bar{F}_n \ F_{n+1}]^T \in \mathbb{R}^{n+1}$ modified flat output vector and $\mathbf{u} = [\boldsymbol{\tau}^T \ u_\lambda \ u_{\phi_0}]^T \in \mathbb{R}^{n+1}$ as the inputs can be expressed as

$$\dot{\mathbf{F}} = \mathbf{f}_F + \bar{\mathbf{g}}_F \mathbf{u}_F, \tag{5.49}$$

where

$$\mathbf{f}_F = \begin{bmatrix} \mathbf{DM}^{-1}\left(\mathbf{W}\dot{\boldsymbol{\theta}}^2 + l\mathbf{SC}_\theta^T \mathbf{f}_R\right) - \boldsymbol{\Phi}''\dot{\lambda}^2 + \mathbf{K}_t \dot{\mathbf{F}}_t \\ \psi_1 - \ddot{\theta}_{ref} + K_n \dot{F}_n \\ K_v(f_2 - \dot{v}_{ref}) \end{bmatrix},$$

$$\mathbf{g}_F = \begin{bmatrix} (\mathbf{DM}^{-1}\mathbf{D}^T) & -\mathbf{b}_1 & -\boldsymbol{\Phi}' \\ \mathbf{0}_{1\times(n-1)} & \psi_3 & \psi_2 \\ \mathbf{0}_{1\times(n-1)} & 0 & 1 \end{bmatrix}.$$

Partitioning the parameter and state-dependent terms of the flat system (5.49) into nominal and uncertain parts as

$$\bar{\mathbf{g}}_F = \hat{\mathbf{g}}_F + \tilde{\mathbf{g}}_F,$$
$$\mathbf{f}_F = \hat{\mathbf{f}}_F + \tilde{\mathbf{f}}_F,$$

one can write

$$\hat{\mathbf{g}}_F \dot{\mathbf{F}} = \hat{\mathbf{g}}_F \hat{\mathbf{f}}_F + \mathbf{u}_F + \mathbf{d}, \tag{5.50}$$

where $\mathbf{d} = \tilde{\mathbf{g}}_F \hat{\mathbf{f}}_F + \hat{\mathbf{g}}_F \tilde{\mathbf{f}}_F + \tilde{\mathbf{g}}_F \tilde{\mathbf{f}}_F - \tilde{\mathbf{g}}_F \dot{\mathbf{F}}$ represents the uncertainty in the flat system.

Assumption 5.2 The uncertainties in the flat system (5.50) has been assumed to be bounded by a constant, i.e.

$$\|\mathbf{d}\| < \rho, \tag{5.51}$$

for $\rho > 0$ being a positive constant.

A robust control law for the flat system (5.50) with uncertainties satisfying Assumption 5.2 can design as

$$\mathbf{u}_F = -\hat{\mathbf{g}}_F \hat{\mathbf{f}}_F - \mathbf{K}_F \mathbf{F} - \eta_0 \operatorname{sat}(\mathbf{F}), \tag{5.52}$$

where $\mathbf{K}_F > 0 \in \mathbb{R}^{(n+1)\times(n+1)}$ is a suitable positive definite feedback gain matrix to be chosen, whereas η_0 is a positive switching gain to be designed to achieve

5.5 Adaptive Robust Control Design for Flat Systems

robustness. The saturation function employed to induce the switching is represented as

$$\text{sat}(\mathbf{F}) = \begin{cases} \frac{\mathbf{F}}{\|\mathbf{F}\|} & \text{for } \|\mathbf{F}\| \geq \epsilon_F, \\ \frac{\mathbf{F}}{\epsilon_F} & \text{otherwise}. \end{cases} \quad (5.53)$$

5.5.2 Adaptation Law

This section presents an approach to update the robust control gains to achieve desired tracking performance. Selecting a constant switching gain often leads to over or under estimation of the gain. Further, one also would need to know the value of ρ a priori to determine the value of η_0. To solve these issues, a dual-rate adaptation law, previously proposed in Chap. 4 [13] has been employed for the switching gains expressed as

$$\dot{\eta}_0 = \begin{cases} \bar{\eta}_{0,i} \|\mathbf{F}\| & for \ \{\eta_0 > 0, \mathbf{F}^T\dot{\mathbf{F}} > 0\} \ or \ \{\eta_0 \leq 0\} \\ -0.5\bar{\eta}_{0,i} \|\mathbf{F}\| & for \ \{\eta_0 > 0, \mathbf{F}^T\dot{\mathbf{F}} \leq 0\} \end{cases}, \quad (5.54)$$

where

$$i = \begin{cases} 1 & for \ \|\mathbf{F}\| \leq \delta \\ 2 & for \ \|\mathbf{F}\| > \delta \end{cases}, \text{ and } \eta_0(t_0) > 0.$$

The parameter δ represents the vicinity of the equilibrium point, whereas $\bar{\eta}_{0,2} > \bar{\eta}_{0,1} > 0$ are the dual-gains of the adaptation law dictating the rate of change of the switching gain η_0. The adaptation law has been designed to update the gains at two different rates depending upon the position of the flat output with respect to the equilibrium point, i.e. if the flat outputs are within the ball of radius δ signifying the vicinity of the equilibrium point, the gain will be updated at a slower rate, otherwise, the gain will be updated at a higher rate. Also, one can observe that the rate of decreasing the gain has been kept as half of the increasing rate to avoid rapid decrement in the gain which may violate the stability criterion and result in rapid increment of the gain.

The closed-loop system with saturated control law (5.52) in the flat system (5.50) can be written as

$$\hat{\mathbf{g}}_F \dot{\mathbf{F}} = -\mathbf{K}_F \mathbf{F} - \eta_0 \text{sat}(\mathbf{F}) + \mathbf{d}. \quad (5.55)$$

The stability of the closed-loop system will be analyzed in the subsequent section to study the efficiency of the proposed control law.

5.5.3 Stability Analysis

Theorem 5.1 *The snake robot system (5.36) remains UUB stable through the application of differential-flatness-based robust control law (5.52) and switching gain adaptation law (5.54).*

Proof Considering a Lyapunov function of the form

$$V = \frac{1}{2}\mathbf{F}^T \hat{\mathbf{g}}_F \mathbf{F} + \frac{1}{2\gamma_0}(\eta_0 - \eta_0^*)^2, \tag{5.56}$$

where η_0^* represent the upper bound on the switching gain. Taking time derivative of the Lyapunov function (5.56) with respect to time, one obtains

$$\dot{V} = \mathbf{F}^T \hat{\mathbf{g}}_F \dot{\mathbf{F}} + \frac{1}{\gamma_0}(\eta_0 - \eta_0^*)\dot{\eta}_0. \tag{5.57}$$

Utilizing the closed-loop system (5.55) in (5.57) yields

$$\dot{V} = -\mathbf{F}^T \mathbf{K}_F \mathbf{F} - \mathbf{F}^T \eta_0 \, \text{sat}(\mathbf{F}) + \mathbf{F}^T \mathbf{d} + \frac{1}{\gamma_0}(\eta_0 - \eta_0^*)\dot{\eta}_0. \tag{5.58}$$

Employing the identity $\mathbf{F}^T \text{sat}(\mathbf{F}) \leq \|\mathbf{F}\|$, (5.58) can be written as

$$\dot{V} = -\mathbf{F}^T \mathbf{K}_F \mathbf{F} - \eta_0 \|\mathbf{F}\| + \mathbf{F}^T \mathbf{d} + \frac{1}{\gamma_0}(\eta_0 - \eta_0^*)\dot{\eta}_0. \tag{5.59}$$

Taking two-norm on both sides of (5.59)

$$\dot{V} \leq -\lambda_{\min}(\mathbf{K}_F) \|\mathbf{F}\|^2 - \eta_0 \|\mathbf{F}\| + \|\mathbf{F}\| \|\mathbf{d}\| + \frac{1}{\gamma_0}(\eta_0 - \eta_0^*)\dot{\eta}_0,$$

$$\dot{V} \leq -\lambda_{\min}(\mathbf{K}_F) \|\mathbf{F}\|^2 - (\eta_0 - \rho) \|\mathbf{F}\| + \frac{1}{\gamma_0}(\eta_0 - \eta_0^*)\dot{\eta}_0. \tag{5.60}$$

Case: 1 $\{\|\mathbf{F}\| \leq \delta\}$
In this case, the flat outputs are already bounded inside the ball of radius δ around the equilibrium point. This doesn't require any further analysis.

Case: 2 $\{\|\mathbf{F}\| > \delta, \mathbf{F}^T \dot{\mathbf{F}} > 0\}$
In this case, the flat outputs are outside the ball of radius δ around the equilibrium point. Employing the adaptation law (5.54) in (5.60), one can write

$$\dot{V} \leq -\lambda_{\min}(\mathbf{K}_F) \|\mathbf{F}\|^2 - (\eta_0 - \rho) \|\mathbf{F}\| + \frac{1}{\gamma_0}(\eta_0 - \eta_0^*)\eta_{0,2} \|\mathbf{F}\|. \tag{5.61}$$

5.5 Adaptive Robust Control Design for Flat Systems

Considering $\gamma_0 = \eta_{0,2}$, (5.61) can be written as

$$\dot{V} \leq -\lambda_{\min}(\mathbf{K}_F) \|\mathbf{F}\|^2 - (\eta_0 - \rho) \|\mathbf{F}\| + (\eta_0 - \eta_0^*) \|\mathbf{F}\|,$$
$$\Rightarrow \dot{V} \leq -\lambda_{\min}(\mathbf{K}_F) \|\mathbf{F}\|^2 - (\eta_0^* - \rho) \|\mathbf{F}\|. \tag{5.62}$$

Hence, one can ensure that $\dot{V} < -\gamma \|\mathbf{F}\|^2$ for $\gamma > 0$ by

$$-\{\lambda_{\min}(\mathbf{K}_F) - \gamma\} \|\mathbf{F}\|^2 - (\eta_0^* - \rho) \|\mathbf{F}\|\} > 0,$$
$$\|\mathbf{F}\| \geq \frac{(\rho - \eta_0^*)}{\{\lambda_{\min}(\mathbf{K}_F) - \gamma\}}. \tag{5.63}$$

This proves UUB stability for Case: 2.
Case: 3 $\{\|\mathbf{F}\| > \delta, \mathbf{F}^T \dot{\mathbf{F}} \leq 0\}$
Similar to Case: 1, employing the adaptation law (5.54) in (5.60), one can write

$$\dot{V} \leq -\lambda_{\min}(\mathbf{K}_F) \|\mathbf{F}\|^2 - (\eta_0 - \rho) \|\mathbf{F}\| - (\eta_0 - \eta_0^*) \|\mathbf{F}\|,$$
$$\Rightarrow \dot{V} \leq -\lambda_{\min}(\mathbf{K}_F) \|\mathbf{F}\|^2 - \{2\eta_0 - \eta_0^* - \rho\} \|\mathbf{F}\|. \tag{5.64}$$

To prove $\dot{V} < -\gamma \|\mathbf{F}\|^2$ for $\gamma > 0$, one has to ensure

$$-\{\lambda_{\min}(\mathbf{K}_F) - \gamma\} \|\mathbf{F}\|^2 - (2\eta_0 - \eta_0^* - \rho) \|\mathbf{F}\|\} > 0,$$
$$\|\mathbf{F}\| \geq \frac{\eta_0^* + \rho - 2\eta_0}{\{\lambda_{\min}(\mathbf{K}_F) - \gamma\}}. \tag{5.65}$$

This proves UUB stability for Case: 3.
Hence, this concludes the proof for Theorem 5.1. □

5.5.4 Simulation Results

The performance of the proposed flatness-based robust control law has been studied and compared to that of the feedback control approach proposed in Sect. 5.4 considering uncertainties in the friction coefficients as detailed in Chap. 2. The physical parameters of the snake robot are same as described in Table 2.1 and trajectory specifications along with the nominal friction coefficients are presented in Table 5.1. The feedback controller gains for the proposed robust control law as well as the feedback control law are given in Table 5.3 and that for the adaptive switching law are provided in Table 5.4.

Table 5.3 Control gains

Parameter	Numerical value
K_{a_1}	20
K_{a_2}	20
K_{n_1}	20
K_{n_2}	20
K_{n+1}	50
K_v	20

Table 5.4 Adaptation law parameters

Parameter	Numerical value
ϵ_F	0.01
δ	0.01
$\eta_0(0)$	3
$\eta_{0,1}$	0.1
$\eta_{0,2}$	0.3

5.5.5 Discussion

The trajectory of the robot CM utilizing the proposed flatness-based robust control methodology in comparison to the feedback control approach presented in Sect. 5.4, has been shown in Fig. 5.18. The proposed approach proves to be more efficient

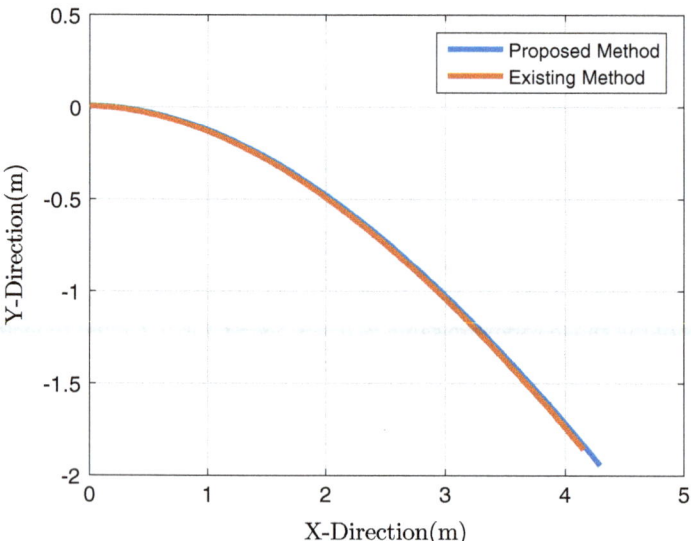

Fig. 5.18 Global trajectory

5.5 Adaptive Robust Control Design for Flat Systems

in providing superior trajectory tracking performance. The superior tracking performance is further confirmed by the tangential velocity error in Fig. 5.19 and the head-angle error in Fig. 5.20. The norm of the torque input required for tracking the trajectory given in Fig. 5.21 shows that the robust control law requires lower control energy compared to the feedback law. The gait function frequency and offset are shown in Fig. 5.22 and Fig. 5.23 respectively. The norm of the actuated flat outputs,

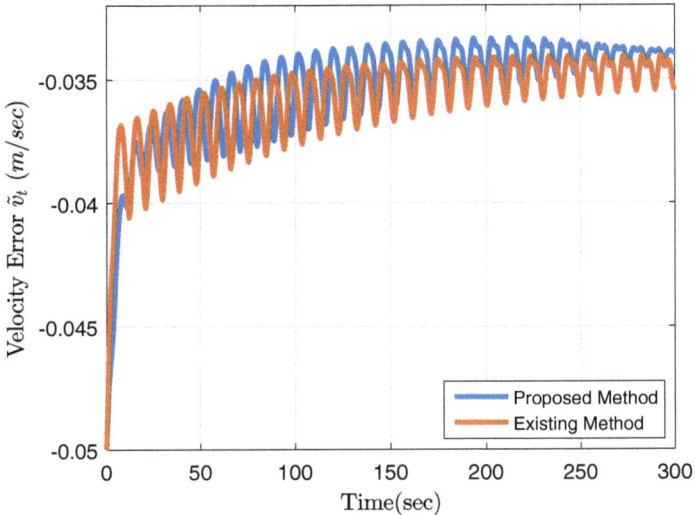

Fig. 5.19 Tangential velocity error

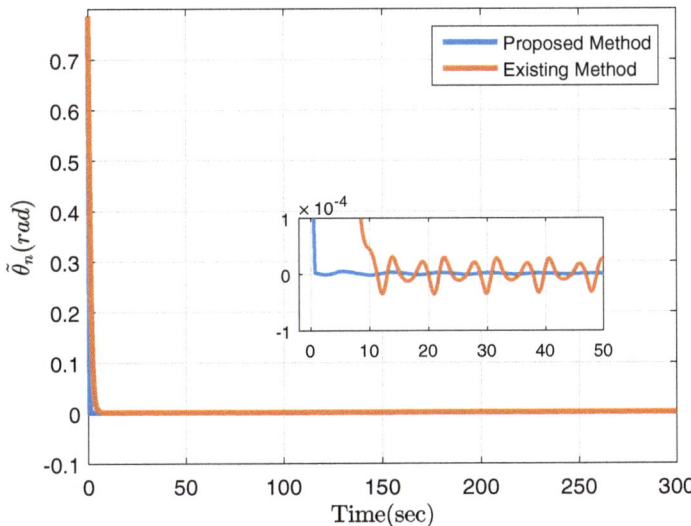

Fig. 5.20 Head-angle error

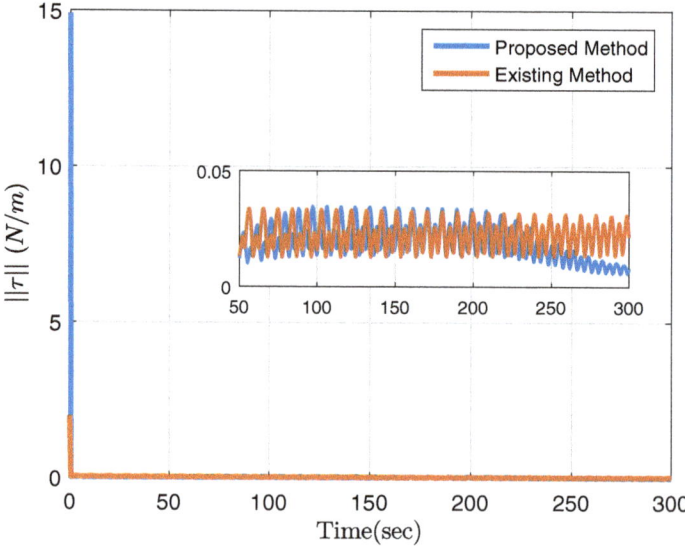

Fig. 5.21 Norm of control effort

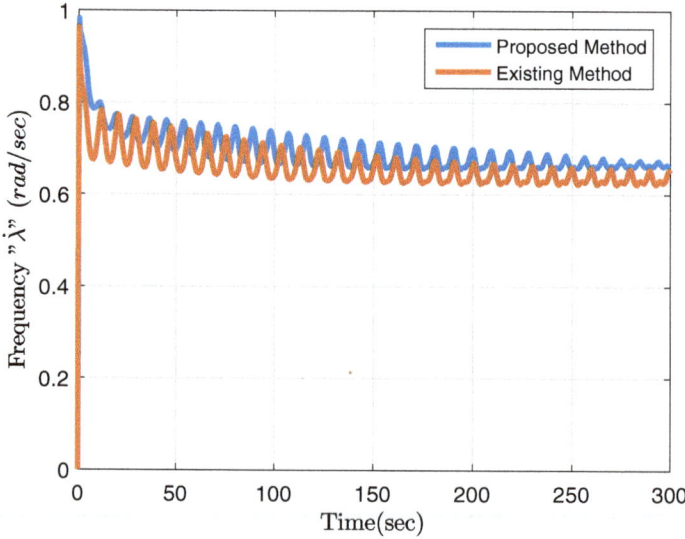

Fig. 5.22 Gait function frequency

i.e. the VHCs are presented in Fig. 5.24 that shows improved convergence to the equilibrium point. The variation in the adaptive switching that remains bounded is shown in Fig. 5.25.

5.5 Adaptive Robust Control Design for Flat Systems

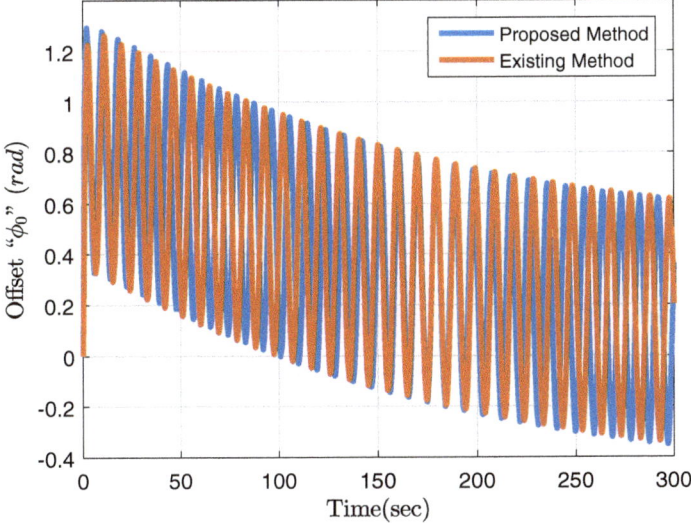

Fig. 5.23 Gait function offset

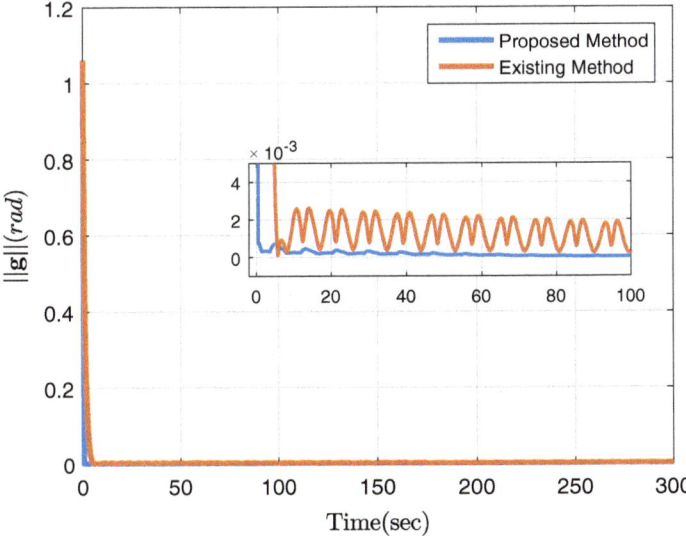

Fig. 5.24 Norm of actuated flat O/p

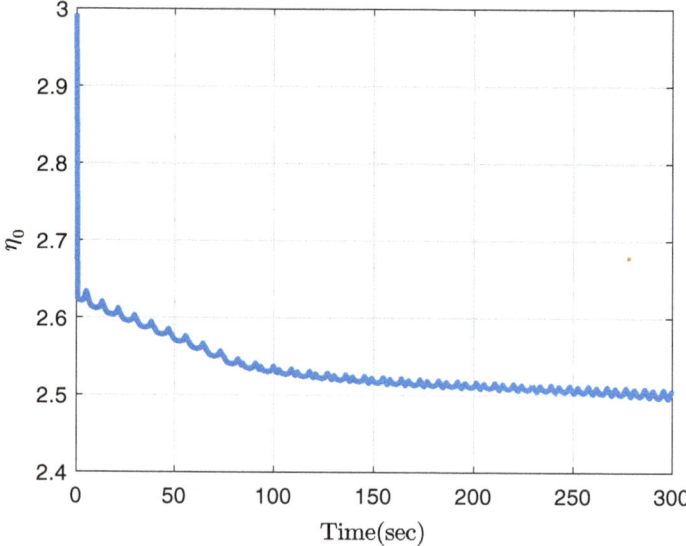

Fig. 5.25 Switching gain adaptation

5.6 Summary

This chapter introduces preliminaries of differential flatness and how it can be used to simplify controller design and path planning. Flatness-based controller design of a wheeled mobile robot has been presented to discuss the pros and cons of this approach. Further, a generalized formulation for flatness-based controller design for a snake robot has been detailed in the chapter. An approach that may lead to the establishment of flatness property for the snake robot dynamics to map the control inputs directly to the flat output have been investigated. Trajectory design in terms of flat outputs would simplify the trajectory tracking problem of the snake robots many fold. However, such an endeavor became intractable due to the nonlinear structure of the snake robot system. Hence, a more restrictive approach has been proposed to obtain a flat system for the snake robot utilizing the serpenoid gait function. A feedback controller design for the same has been demonstrated and verified through simulations. The proposed methodology exhibits improved heading and velocity tracking with respect to the current state-of-the-art approach for a deterministic model. The flatness-based approach has been employed to design an adaptive robust control law to achieve robustness towards uncertainties. The flatness-based approach will be utilized in future to design superior controllers capable of handling real world uncertainties and incorporate input saturation. The flatness-based methodology will provide a simplistic framework to address such practical issues in a convenient way.

References

1. Fliess, M., Lévine, J., Martin, P., Rouchon, P.: Flatness and defect of non-linear systems: introductory theory and examples. Int. J. Control **61**(6), 1327–1361 (1995)
2. Lévine, J.: On necessary and sufficient conditions for differential flatness. Appl. Algebra Eng. Commun. Comput. **22**(1), 47–90 (2011)
3. Nicolau, F., Respondek, W.: Multi-input control-affine systems linearizable via one-fold prolongation and their flatness. In: 2013 IEEE 52nd Annual Conference on Decision and Control (CDC), pp. 3249–3254. IEEE (2013)
4. Ryu, J.C., Agrawal, S.K.: Differential flatness-based robust control of mobile robots in the presence of slip. Int. J. Robot. Res. **30**(4), 463–475 (2011)
5. Tang, C.P.: Differential flatness-based kinematic and dynamic control of a differentially driven wheeled mobile robot. In: 2009 IEEE International Conference on Robotics and Biomimetics (ROBIO), pp. 2267–2272. IEEE (2009)
6. Rouchon, P., Fliess, M., Levine, J., Martin, P.: Flatness and motion planning: the car with n trailers. In: Proceedings of the ECC'93, Groningen, pp. 1518–1522 (1993)
7. Delaleau, E., Louis, J.P., Ortega, R.: Modeling and control of induction motors. Int. J. Appl. Math. Comput. Sci. **11**, 105–129 (2001)
8. Agrawal, S.K., Sangwan, V.: Differentially flat designs of underactuated open-chain planar robots. IEEE Trans. Robot. **24**(6), 1445–1451 (2008)
9. Mukherjee, J., Kar, I.N., Mukherjee, S.: Kinematic control of wheeled mobile robot: an error based differentially flat system approach. In: 2015 Annual IEEE India Conference (INDICON), pp. 1–6 (2015). https://doi.org/10.1109/INDICON.2015.7443561
10. Blažič, S.: A novel trajectory-tracking control law for wheeled mobile robots. Robot. Auton. Syst. **59**(11), 1001–1007 (2011)
11. Kanayama, Y., Kimura, Y., Miyazaki, F., Noguchi, T.: A stable tracking control method for an autonomous mobile robot (1990)
12. Mohammadi, A., Rezapour, E., Maggiore, M., Pettersen, K.Y.: Maneuvering control of planar snake robots using virtual holonomic constraints. IEEE Trans. Control Syst. Technol. **24**(3), 884–899 (2016). https://doi.org/10.1109/TCST.2015.2467208
13. Mukherjee, J., Roy, S., Kar, I.N., Mukherjee, S.: Maneuvering control of planar snake robot: an adaptive robust approach with artificial time delay. Int. J. Robust Nonlinear Control (2021). https://doi.org/10.1002/rnc.5430 [In Press]

Chapter 6
Modeling of In-Pipe Snake Robot Motion

Abstract The control laws presented in Chaps. 2, 3 and 4 were aimed at achieving improved tracking performance utilizing least control effort for the planar snake robot model. This particular form of locomotion for a snake robot is inspired by the commonest mode of snake motion called lateral undulation. Other modes of generating propagating force for a snake robot, in channel or around a cylinder for example, have not been extensively studied. This chapter is targeted to establish and obtain a mathematical model of a snake robot inside a constrained space like a pipe or a channel. The model has been employed to design control laws for the robot's efficient motion through the pipe. To address uncertainties that may occur during such a motion, a flatness-based adaptive robust control approach presented in Chap. 5 has been employed as well.

Pipes and channels are integral part of modern-day industry and infrastructure where these are often employed to carry poisonous gases, water, oil, hot steam and many other essential fluids. Pipes also house power and communication cables in industries and also in long distance operations. Hence, inspection of these pipes over time is essential toward the proper functioning of such industries. In the near future, we expect to see an effective solution of these inspection problems through autonomous robots. Motion characteristics of the robots inside such channels enable deployment in unstructured environments as well like natural tunnels or debris in case of a collapsed structure. The various challenges to be addressed in this work include channel geometry, pipe orientation, physical condition of the pipe walls, etc. as these will directly influence the motion of any kind of robot that is inserted.

A modular robot, each module having 2-DoFs, has been presented in [1], whereas [2] have proposed a robot with active wheeled units at front and rear with the passive elements in intermediate joints. To achieve greater traction force, [3] have designed and tested a robot with six modular arms, each having six wheels with an angular separation of 60°. A multi-vehicle robot has been proposed in [4] where each module is designed for a specific job like traction, ultrasonic test, rotation, sensor measurement, camera visual, etc. A three-wheel robot with the wheels mounted on a scissor structure pressing on the pipe wall by means of a spring with levers (MOGRER) to achieve in-pipe movement has been detailed in [5], while [6] presents a modular robot, specifically designed to enable motion through pipes with elbow bends and other configurations. A four-wheel-driven robot with a navigation algorithm based on ego-pose estimation using an omnidirectional camera and laser modules has been discussed in [7]. Also, [8] have designed a modular robot with separate steering, driving, camera module and a functional module to generate the required traction. A conceptualization of a snake robot for pipe inspection has been presented in [9, 10].

Being able to model the motion of a snake robot inside constrained spaces is a precursor to analyze and design of robotic snakes for the targeted applications. A methodical approach using Newton–Euler method has been presented in this chapter to formulate the dynamics of snake robot inside a channel. The free-body diagram of a single link has been derived followed by equating forces along the global X- and Y-direction and moments about the link CM. Hertzian theory of contact has been utilized to model the contact force generated by each link when in contact with the walls on either side. A continuous-time saturation function has been employed to model the contact make and break phenomenon during the motion. A Coulomb friction model has been used to access the traction force generated during the link-wall contact, which is the primary contributor to the propagating force of the overall snake robot. Due to the fact that these forces doesn't necessarily act through the link CM, there contributions towards the moment of the link has been computed as well. The forces computed for individual links have been cascaded to define the dynamic equation of motion for the snake robot inside the channel. A serpenoid gait function specific to the channel and link dimension, keeping in mind that making wall contact is crucial for propagation, has been defined. The control law presented in Chap. 1 has been used to demonstrate tracking of the robot inside the channel.

The rest of the chapter is organized as follows: The mathematical model and its various components for the constrained motion of a snake robot has been proposed in Sect. 6.1; The control objective and the conventional control framework with simulation results has been discussed in Sect. 6.2; Flatness-based adaptive robust controller has been presented with simulation results in Sect. 6.3 to address uncertainties; Finally, the chapter has been summarized in Sect. 6.4.

6.1 Dynamic Modeling

In this section, the specific body shape required for the robot to execute locomotion through the channel has been detailed. Further, the dynamic equations of motion for the whole robot have been derived by equating forces and moments of each link using Newton–Euler method.

The snake robot requires support and friction force to move forward inside a pipe or a channel. It has been assumed that the robot makes contact with the pipe walls as shown in Fig. 6.1 and generates propagating force to move ahead. Maintaining adequate number of contact is important because it fixes point on the robot in a particular position and also generates the forward propagating force. Each link transits from

6.1 Dynamic Modeling

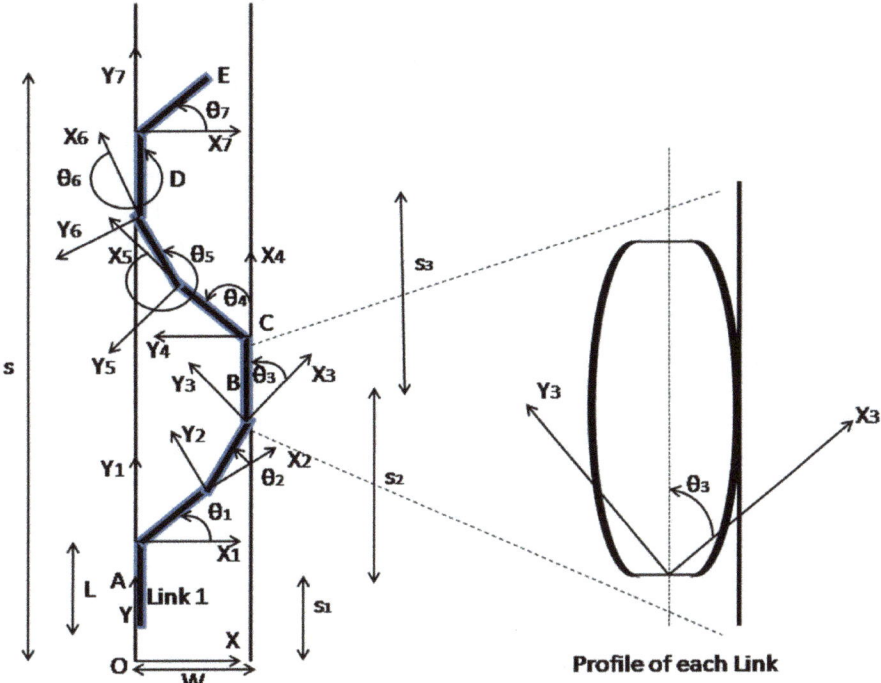

Fig. 6.1 Schematic diagram of snake locomotion inside channel

one wall to the other due to the serpenoid gait. All these forces are to be coordinated to achieve overall locomotion of the robot. The shape of each link has been chosen to be elliptical with a being the semi-minor and b being the semi-major axis. The kinematic equations for the snake robot can be derived from (1.5) with the half-link length being $l = b$. The following subsection details the contact force model adopted for dynamic modeling.

6.1.1 Contact Force Model

As has already been discussed, it is essential for the links of the snake robot to establish contact with the channel walls to generate adequate propagating force. Contact is implemented through mutual *Penetration*, i.e. deformation of the bodies in contact. The penetration extent is used to compute contact force. For elliptical link profile, the polar coordinate of the link surface is given as

$$x_{i,\psi} = x_i + b \cos \psi \cos \theta_i - a \sin \psi \sin \theta_i, \tag{6.1a}$$

$$y_{i,\psi} = y_i + b \cos \psi \sin \theta_i + a \sin \psi \cos \theta_i, \tag{6.1b}$$

where ψ is the polar angle for each point over the link surface. It is to be noted that links transit from one wall to the other during the course of motion. Hence, contact make and break between the link and the wall has to be incorporated in to contact model. The free-body diagram of a link in contact with a wall has been shown in Fig. 6.2. The maximum penetration for a link is obtained by computing the coordinate of the point on the link with minimum or maximum abscissa. Differentiating $x_{i,\psi}$ (6.1a) with respect to ψ and equating it to zero, yields the extremums. Corresponding values of ψ are utilized to obtain the coordinates (6.1) of these points for a specific link angle expressed as

$$\begin{aligned} x_{i,min} &= x_i - \sqrt{(b^2 cos^2\theta_i + a^2 sin^2\theta_i)}, \\ y_{x_{i,min}} &= y_i + \frac{(a^2 - b^2) \cos\theta_i \sin\theta_i}{\sqrt{(b^2 cos^2\theta_i + a^2 sin^2\theta_i)}}, \end{aligned} \qquad (6.2)$$

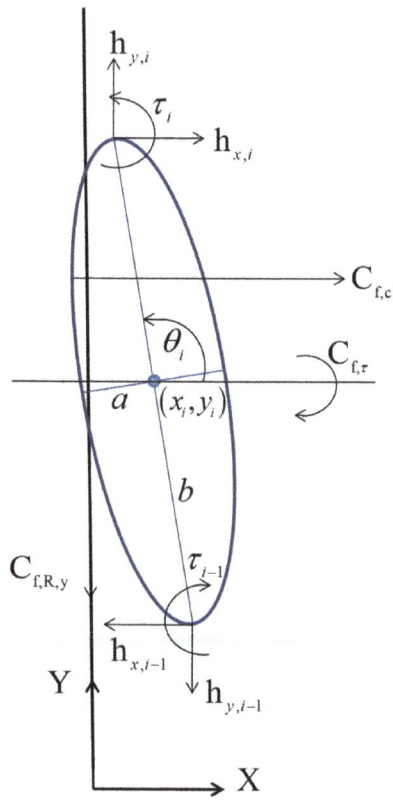

Fig. 6.2 Free-body diagram of a link in contact with a wall

6.1 Dynamic Modeling

$$x_{i,max} = x_i + \sqrt{(b^2\cos^2\theta_i + a^2\sin^2\theta_i)},$$
$$y_{x_{i,max}} = y_i - \frac{(a^2 - b^2)\cos\theta_i \sin\theta_i}{\sqrt{(b^2\cos^2\theta_i + a^2\sin^2\theta_i)}}. \tag{6.3}$$

The contact force model based on Hertz theory for elastic bodies with smooth geometry can be expressed as [11]

$$\delta_p = \left(\frac{9P^2}{16RE_r^2}\right)^{1/3}, \tag{6.4}$$

where δ is the normal penetration, P is the applied force, R is the radius of curvature of the curved surface and E_r is given by

$$\frac{1}{E_r} = \left(\frac{1-v_1^2}{E_1} - \frac{1-v_2^2}{E_2}\right), \tag{6.5}$$

where v_i and E_i for $i = 1, 2$, are the Poisson's ratio and Modulus of elasticity of the two bodies, respectively. The link surface is considered to be made of compliant material, whereas the channel wall is assumed to be rigid. Hence, the deformation only of the link surface E_r can be computed as

$$E_r = \frac{E_{rub}}{1 - v_{rub}}, \tag{6.6}$$

where E_{rub} and v_{rub} represents the modulus of elasticity and Poisson's ratio of the link surface, respectively. Thus, the contact force can be derived from (6.4) for penetration δ and can be expressed as

$$P = \left(\frac{16 E_{rub} R}{9(1 - v_{rub}^2)}\right)^{1/2} \delta^{3/2}. \tag{6.7}$$

The above expression holds true when the link is in contact with the pipe wall. However, to incorporate contact make and break, the expression for penetration has been modified using a *Logistic Function* for the left and the right wall as

$$\delta_l = \frac{\sqrt{x_{i,min}^2}}{2}\left(1 + \tanh\left(k_{con}(x_{i,min} + \epsilon_{rub})\right)\right), \tag{6.8}$$

$$\delta_r = \frac{\sqrt{(d - x_{i,max})^2}}{2}\left(1 + \tanh\left(k_{con}(d + \epsilon_{rub} - x_{i,max})\right)\right), \tag{6.9}$$

where k_{con} is a positive scalar constant that modifies the slope of the function and ϵ_{rub} is the maximum virtual penetration. Here δ_l represents the penetration at the left wall, whereas δ_r signifies the same at the right wall. In general, contact make and break is a discontinuous phenomenon. However, in this work, the inclusion of the differentiable saturation function makes this smooth and continuous. The contact forces (6.7) due to the respective walls can be expressed as

$$N_{i,l} = \left(\frac{16E_{rub}R}{9(1-v_{rub}^2)}\right)^{1/2} \delta_l^{3/2}, \quad (6.10)$$

$$N_{i,r} = -\left(\frac{16E_{rub}R}{9(1-v_{rub}^2)}\right)^{1/2} \delta_r^{3/2}.$$

The contact forces acting on a link due to the left and the right wall with respect to its CM position are shown in Figs. 6.3 and 6.4. Contact forces shown in the aforementioned figures have been computed using radius of curvature as $R = b^2/a$ for the contact point $(\pm a, 0)$. The physical parameters of the link and the channel are given in Table 6.1. The shift from contact to no contact happens over 1 mm, whereas the link transits from one wall to the other without any effective contact force over 110 mm.

A contact forces acting on each links by virtue of the left and the right wall can be cascaded in vector form as $\mathbf{N}_l = [N_{1,l}\ N_{2,l}\ \ldots\ N_{n,l}]^T$ and $\mathbf{N}_r = [N_{1,r}\ N_{2,r}\ \ldots\ N_{n,r}]^T$. The vector representing the total contact force acting on each link of the robot can be computed as

$$\mathbf{C}_{f,c} = \mathbf{N}_l + \mathbf{N}_r. \quad (6.11)$$

The following subsection details the approach adopted to model the traction force at the contact points.

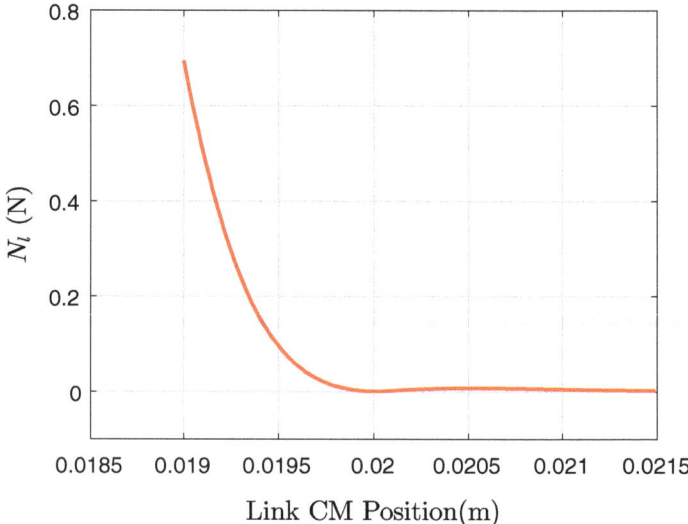

Fig. 6.3 Left wall contact force

6.1 Dynamic Modeling

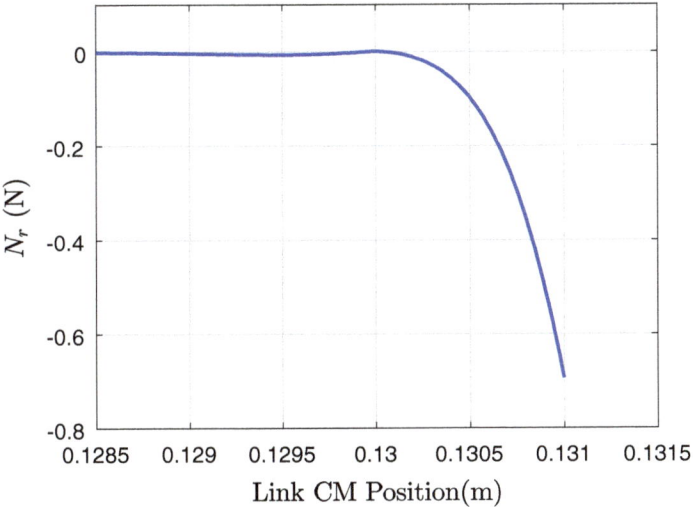

Fig. 6.4 Right wall contact force

6.1.2 Friction Force Model

A Coulomb friction model has been considered in this work to determine the traction force generated at the point of contact between the link and the channel wall. The velocity of the contact point along the wall is used to switch the direction of the friction force using a logistic function. The contact point velocity can be determined as

$$v_{i,l} = (\dot{y}_i - r_{i,l}\,\dot{\theta}_i),$$
$$v_{i,r} = (\dot{y}_i + r_{i,r}\,\dot{\theta}_i), \qquad (6.12)$$

where $\dot{y}_i$ is the linear velocity of the ith link along the pipe. Also, $r_{i,0}$ and $r_{i,d}$ represent the distance between the link CM and its corresponding contact points expressed as

$$r_{i,l} = \sqrt{x_i^2 + (y_{x_{i,min}} - y_i)^2},$$
$$r_{i,r} = \sqrt{(d - x_i)^2 + (y_{x_{i,max}} - y_i)^2}. \qquad (6.13)$$

The coefficient of friction can be computed using a saturation function of the contact point velocity as

$$\mu_{i,l} = -\mu_{fric}\big(1 - \tanh(k_{fric}\, v_{i,l})\big),$$
$$\mu_{i,r} = -\mu_{fric}\big(1 - \tanh(k_{fric}\, v_{i,r})\big). \qquad (6.14)$$

Subsequently, the friction force acting on a particular link corresponding to each wall is computed through a Coulomb friction model as the product of the friction coefficient (6.14) and the contact force (6.10) as

$$C_{i,l} = -\mu_{i,l} N_{i,l},$$
$$C_{i,r} = \mu_{i,r} N_{i,r}. \quad (6.15)$$

A vector representing the friction force acting on each link of the snake robot along the wall can be computed from as

$$\mathbf{C}_{f,y} = \mathbf{C}_{f,l} + \mathbf{C}_{f,r}, \quad (6.16)$$

where $\mathbf{C}_{f,l} = [C_{1,l}\ C_{2,l}\ \ldots\ C_{n,l}]^T$ and $\mathbf{C}_{f,r} = [C_{1,r}\ C_{2,r}\ \ldots\ C_{n,r}]^T$ are the friction forces corresponding to each wall acting on each link cascaded in a vector form.

In the classical snake robot analysis [12, 13], forces are generated through anisotropic friction between the ground and the links. However, for motion inside the channel, the contact and friction forces have been modeled to be generated only on the links which are in contact with the wall. These forces induce a moment on the link in contact, which has been discussed in the following subsection.

6.1.3 Moment Due to Contact and Friction Forces

The normal and traction forces generated due to the interaction of a link and the channel wall, as shown in Fig. 6.2, do not pass through the CM and in general would impart a moment. The moment resulting from normal forces about each link CM can be computed using the *cross-product* as

$$\mathbf{C}_{f,c,\tau,0,i} = \left[(x_{i,min} - x_i)\ (y_{x_{i,min}} - y_i)\ 0\right]^T \times \left[N_{0,i}\ 0\ 0\right]^T,$$
$$\mathbf{C}_{f,c,\tau,d,i} = \left[(x_{i,max} - x_i)\ (y_{x_{i,max}} - y_i)\ 0\right]^T \times \left[N_{d,i}\ 0\ 0\right]^T. \quad (6.17)$$

Similarly, the moment from the friction forces for a particular link can be expressed as

$$\mathbf{C}_{f,y,\tau,0,i} = \left[-x_i\ 0\ 0\right]^T \times \left[0\ C_{fric,i,0}\ 0\right]^T,$$
$$\mathbf{C}_{f,y,\tau,d,i} = \left[(d - x_i)\ 0\ 0\right]^T \times \left[0\ C_{fric,i,d}\ 0\right]^T. \quad (6.18)$$

The total moment acting on a link can be computed by adding the components due to the contact force (6.17) and the friction force (6.18) at both the wall as

$$C_{f,\tau,i} = \begin{bmatrix} 0 & 0 & 1 \end{bmatrix} (\mathbf{C}_{f,c,\tau,0,i} + \mathbf{C}_{f,c,\tau,d,i} + \mathbf{C}_{f,y,\tau,0,i} + \mathbf{C}_{f,y,\tau,d,i}). \quad (6.19)$$

A vector representing the moment acting on each link of the robot can be concatenated as $\mathbf{C}_{f,\tau} = [C_{f,\tau,1} \ C_{f,\tau,2} \ \ldots \ C_{f,\tau,n}]^T$.

6.1.4 Dynamic Equations

A Newton–Euler formulation has been employed to derive a mathematical expression for the motion of a snake robot moving within a channel. In this approach, the forces and torques working on each link are compounded to obtain the equations of motion for the whole robot. Equating forces along the global $X-$ and Y-directions for each link we get [12]

$$m \begin{bmatrix} \ddot{\mathbf{x}} \\ \ddot{\mathbf{y}} \end{bmatrix} = \begin{bmatrix} \mathbf{C}_{f,c} \\ \mathbf{C}_{f,y} \end{bmatrix} + \begin{bmatrix} \mathbf{D}^T \mathbf{h}_x \\ \mathbf{D}^T \mathbf{h}_y \end{bmatrix}, \qquad (6.20)$$

where $\mathbf{h}_x \in \mathbf{R}^{n-1}$ and $\mathbf{h}_y \in \mathbf{R}^{n-1}$ are the vectors representing the joint constraint forces generated due to the interaction between consecutive links, along the $X-$ and Y-direction. These forces can be solved from (6.20) as

$$\begin{aligned} \mathbf{h}_x &= (\mathbf{DD}^T)^{-1} \left\{ mb\mathbf{A} \left(\mathbf{C}_\theta \dot{\theta}^2 + \mathbf{S}_\theta \ddot{\theta} \right) - \mathbf{DC}_{f,c} \right\}, \\ \mathbf{h}_y &= (\mathbf{DD}^T)^{-1} \left\{ mb\mathbf{A} \left(\mathbf{S}_\theta \dot{\theta}^2 - \mathbf{C}_\theta \ddot{\theta} \right) - \mathbf{DC}_{f,y} \right\}. \end{aligned} \qquad (6.21)$$

From torque balance for each link, the angular dynamics is given as

$$\mathbf{J}\ddot{\theta} = -b\mathbf{S}_\theta \mathbf{A}^T \mathbf{h}_x + b\mathbf{C}_\theta \mathbf{A}^T \mathbf{h}_y - \mathbf{C}_{f,\tau} - \mu_\tau \dot{\theta} + \mathbf{D}^T \tau. \qquad (6.22)$$

An additional viscous damping proportional to the angular rate has been introduced as an energy dissipation term to the dynamics of each link with μ_τ being the damping coefficient. The internal forces $\mathbf{h}_x$ and $\mathbf{h}_y$ are eliminated from (6.22) using (6.20) and (6.21) to obtain the angular dynamics as

$$\mathbf{M}\ddot{\theta} = \mathbf{W}\dot{\theta}^2 + b\mathbf{S}_\theta \mathbf{NC}_{f,c} - b\mathbf{C}_\theta \mathbf{NC}_{f,y} - \mathbf{C}_{f,\tau} - \mu_\tau \dot{\theta} + \mathbf{D}^T \tau, \qquad (6.23)$$

where

$$\begin{aligned} \mathbf{M} &= J\mathbf{I}_{n \times n} + mb^2 \mathbf{S}_\theta \mathbf{V} \mathbf{S}_\theta + mb^2 \mathbf{C}_\theta \mathbf{V} \mathbf{C}_\theta, \\ \mathbf{W} &= mb^2 \mathbf{C}_\theta \mathbf{V} \mathbf{S}_\theta - mb^2 \mathbf{S}_\theta \mathbf{V} \mathbf{C}_\theta. \end{aligned}$$

Furthermore, the equations of motion of the robot CM can be derived by multiplying (6.20) with $\mathbf{e}^T$ and using the relation $\mathbf{e}^T \mathbf{D}^T = \mathbf{0}$ to obtain

$$\ddot{\mathbf{p}} = \frac{1}{nm} \begin{bmatrix} \mathbf{e}^T \mathbf{C}_{f,c} \\ \mathbf{e}^T \mathbf{C}_{f,y} \end{bmatrix}. \qquad (6.24)$$

Hence, the angular dynamics for all the links (6.23) and translational dynamics of the robot CM (6.24) together describe the equations of motion for the whole snake robot inside a channel with generalized coordinate $\mathbf{q} = [\boldsymbol{\theta}^T \ \mathbf{p}^T]^T$ and generalized torque vector $\boldsymbol{\tau}$.

6.1.5 Serpenoid Gait Function

To achieve desired motion, a serpenoid gait function [14] that results in to a time-varying body shape is applied to the joints of the snake robot. The gait function for motion of a snake robot in a plane has been detailed in Sect. 1.5.1 as

$$\phi_i = \alpha \sin(\omega t + (i-1)\delta), \tag{6.25}$$

where α, ω and δ represent the amplitude, frequency and phase delay. The gait function imposes an wave like body shape on the robot to establish contact on either wall. The values of the gait parameters are selected to obtain desired characteristics for motion in a channel. The amplitude of the gait function α may be chosen according to the physical dimensions of the robot as well as the pipe as

$$\alpha = \sin^{-1}\left\{\frac{(d + 2\epsilon_{rub} - 2a)}{b(w_l - 2)}\right\},$$

where w_l represents the number of the links making a complete wave. The value of α ensures that the snake robot attempts an undulation of a width slightly greater than the pipe width so as to generate contact forces. With w_l number of links in a wave, the relative delay between consecutive joints can be computed by $\delta = 2\pi/w_l$. The other gait parameters are set to solve a tracking problem is to be discussed next.

6.2 Conventional Control Methodology

The control law discussed in Chap. 1 has been employed to achieve a desired motion for the in-channel snake robot system proposed in Sect. 6.1. The control objectives to be addressed can be accounted as

- The joints are to be tracked to the serpenoid gait utilizing a VHC-based approach.
- The head angle of the robot is to track to the orientation of the channel to ensure efficient motion along the same.
- The velocity of the robot along the channel is to be regulated to a reference value to ensure desired motion inside the channel.

6.2.1 Body-Shape Control

The control law described in Sects. 1.5 and 1.6 has been employed in this work to impose the desired gait function on the snake robot while tracking a desired heading and velocity along the channel. With modification in the system dynamics, the control laws employed in the aforementioned sections have been modified accordingly. The VHCs adopted can be expressed as [15]

$$\mathbf{h}(\lambda, \phi_0, \theta) = \mathbf{D}\theta - \mathbf{\Phi}(\lambda) - \mathbf{b}_1\phi_0. \tag{6.26}$$

To regulate the VHCs to zero, i.e. track the joint angles to the gait function (6.25), the feedback control law becomes

$$\tau = (\mathbf{DM}^{-1}\mathbf{D})^{-1}(-\mathbf{DM}^{-1}\mathbf{W}\dot{\theta}^2 - b\mathbf{DM}^{-1}\mathbf{S}_\theta \mathbf{N}\mathbf{C}_{f,c} + b\mathbf{DM}^{-1}\mathbf{S}_\theta \mathbf{N}\mathbf{C}_{f,y}$$
$$+ \mu_\tau \mathbf{DM}^{-1}\dot{\theta} + \mathbf{DM}^{-1}\mathbf{C}_{f,\tau} + \boldsymbol{\phi}''(\lambda)\dot{\lambda}^2 + \boldsymbol{\phi}'(\lambda)\ddot{\lambda} + \mathbf{b}_1\ddot{\phi}_0 - \mathbf{K}_P\mathbf{h} - \mathbf{K}_D\dot{\mathbf{h}}), \tag{6.27}$$

where $\mathbf{K}_P$ and $\mathbf{K}_D$ are positive definite matrices to be chosen of appropriate dimension.

6.2.2 Head-Angle Control

The head and the orientation of a robotic snake remains physically bounded inside a pipe due to the pipe walls. Yet, to achieve an efficient motion maintaining contact with the pipe walls, the head angle has to be tracked to a stable limit cycle along the center line of the pipe. The error in the heading can be expressed as

$$\tilde{\theta}_n = \theta_n - \theta_{n_{ref}}, \tag{6.28}$$

where $\theta_{n_{ref}}$ is the orientation of the channel. The singular-perturbation-based control law can be expressed as

$$u_{\phi_0} = \frac{1}{\psi_3}\left(\frac{1}{\epsilon}(\dot{\tilde{\theta}}_n + k_N\tilde{\theta}_n)\right) - k_1\phi_0 - k_2\dot{\phi}_0, \tag{6.29}$$

where k_N, k_1 and k_2 are positive gains and ϵ_1 is sufficiently small scalar to be chosen. Gait parameter ϕ_0 and its derivative can be obtained from the solution of the compensator given as

$$\ddot{\phi}_0 = u_{\phi_0}. \tag{6.30}$$

6.2.3 Velocity Control

For a snake robot in an open plane, the velocity depends upon the frequency of the serpenoid gait function and the width of the undulation. The undulation of the robot in this case is obviously constrained by the channel width. The tangential velocity of the robot along the channel can be obtained as

$$v_t = u_{\theta_n}^T \dot{p}. \tag{6.31}$$

The error in the tangential velocity can be written as

$$\tilde{v}_t = v_t - v_{ref}, \tag{6.32}$$

where v_{ref} is the desired reference velocity. To track the velocity of the robot along the channel to a desired value, the feedback-linearization-based control law can be described as

$$u_\lambda = -k_z(\dot{\lambda} + k_\lambda \tilde{v}_t) - k_\lambda f_v, \tag{6.33}$$

where k_z and k_λ are positive gains to be chosen and f_v can be computed as

$$f_v = \frac{1}{nm} u_{\theta_n}^T \begin{bmatrix} \mathbf{e}^T \mathbf{C}_{f,c} \\ \mathbf{e}^T \mathbf{C}_{f,y} \end{bmatrix} + v_{\theta_n}^T \dot{p}\dot{\theta}_n. \tag{6.34}$$

The gait frequency $\dot{\lambda}$ is obtained by solving the compensator given as

$$\ddot{\lambda} = u_\lambda. \tag{6.35}$$

The tracking performance of the detailed control law has been verified through simulation results enclosed in the following section.

6.2.4 Simulation Results

This section presents the simulation setup and the results from the snake robot motion inside a channel when controlled by the aforementioned control law. The physical specifications of the system considered for simulation are given in Table 6.1. The simulation was run for 300 s and the initial condition of the generalized coordinates and compensators states are set to $\theta(0) = \dot{\theta}(0) = \mathbf{p}(0) = \dot{\mathbf{p}}(0) = \mathbf{0}$ and $\lambda = \dot{\lambda} = \phi_0 = \dot{\phi}_0 = 0$. The controller gains utilized for the control law discussed in the previous section are shown in Table 6.2.

Table 6.1 System parameters

Simulation parameter	Numerical value
n	10
m	1 Kg
a	0.02 m
b	0.1 m
J	0.0026 Kg m^2
d	0.15 m
E_{rub}	0.05 MPa
v_{rub}	0.49
ϵ_{rub}	0.001 m
μ_{fric}	0.5 Ns/m
μ_τ	0.4 N-m-s/rad
k_{con}	1000
k_{fric}	150

Table 6.2 Controller parameters

Controller parameter	Numerical value
θ_{ref}	$\pi/2$
v_{ref}	0.07 m/s
α	0.35 rad
δ	1.048 rad
k_1	1
k_2	1
K_N	20
ϵ	0.1
K_λ	15
K_Z	30
$\mathbf{K}_P$	10 I_{n-1}
$\mathbf{K}_D$	10 I_{n-1}

6.2.5 Discussion

The global trajectory of the robot CM is shown in Fig. 6.5 which exhibits limit cycle behavior inside the pipe with an amplitude of 13 mm compared to 150 mm channel width. The velocity profile of the snake robot along the global X- and Y-directions shown in Fig. 6.6 shows bounded velocity in the normal direction and velocity tracking along the channel. The CM trajectories for the first and the second link exhibit their transition from one wall to the other by virtue of the gait function to generate normal and traction forces as shown in Figs. 6.7 and 6.8, respectively. The manner in which contacts are being made by the first two links with the left and right

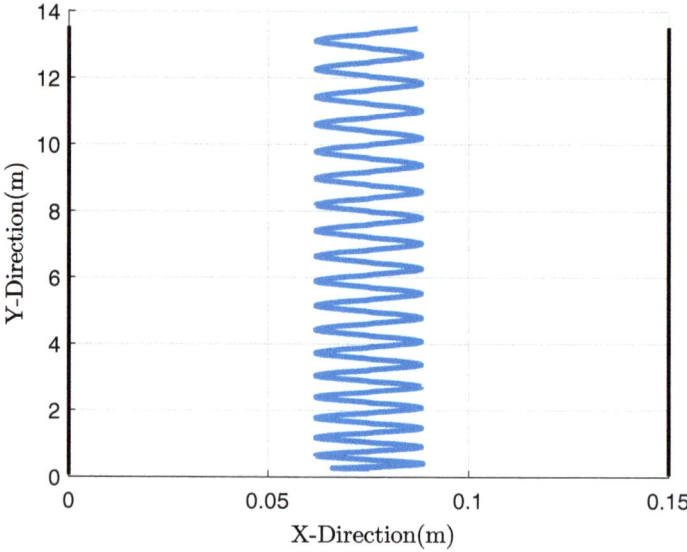

Fig. 6.5 Trajectory of robot CM

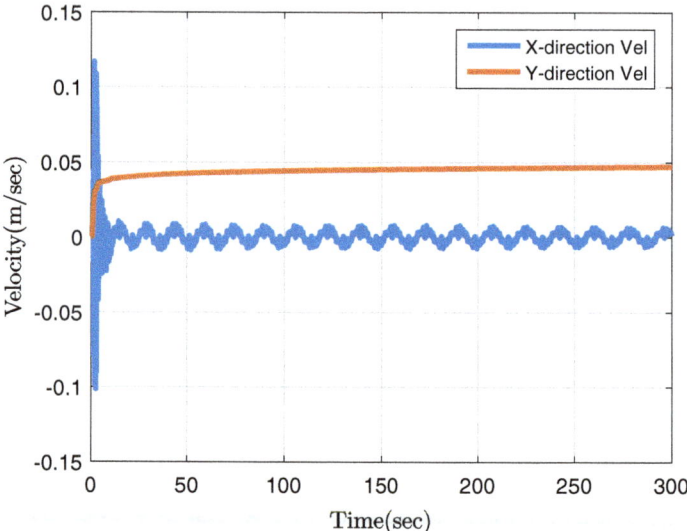

Fig. 6.6 Velocity profile of robot CM

6.2 Conventional Control Methodology

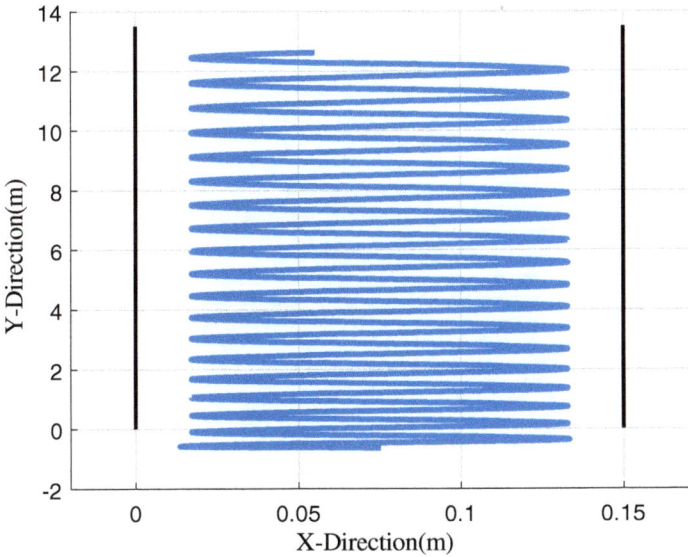

Fig. 6.7 First link CM trajectory

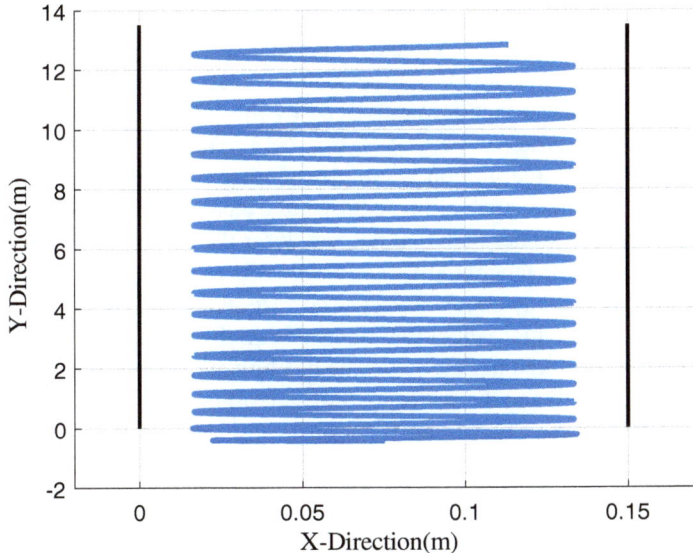

Fig. 6.8 Second link CM trajectory

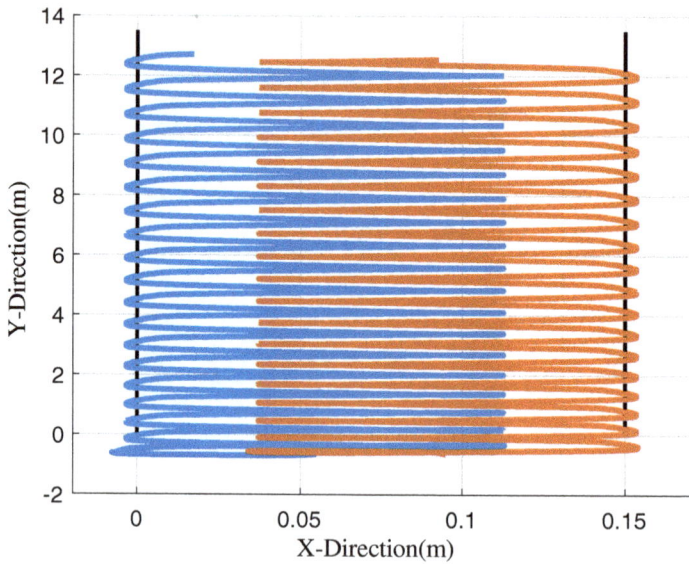

Fig. 6.9 First link contact trajectory

channel wall are shown in Figs. 6.9 and 6.10 respectively. The virtual penetrations responsible for the contact forces can be seen in these figures. The error between joint angles and the gait function represented by the VHCs converge to the origin with accuracy of the order of 10^{-5} as in Fig. 6.11. The norm of the total control effort employed to execute motion while maintaining contact with the side wall is shown in Fig. 6.12. The control effort required to execute motion inside the channel can be observed to be higher than that necessary for the planar motion with ground friction. The error in the tangential velocity shown in Fig. 6.13 demonstrates stable performance of the velocity control law. The corresponding gait frequency needed to attain the desired tangential velocity remains within the practically realizable limit of 2 rad/sec as is exhibited in Fig. 6.14. The heading angle error in Fig. 6.15 shows boundedness about zero and the corresponding gait offset is being shown in Fig. 6.16 which reassures that the axis of gait function is aligned with the centerline of the channel.

6.3 Flatness-Based Adaptive Robust Control

Like any other system, a snake robot inside a channel is also susceptible to uncertainties. Internal parameters like mass, length and inertia of the robot, although being measurable can have minor inaccuracy. Moreover, non-uniformity in the channel walls, variation in orientation and variation in the channel wall condition can induce

6.3 Flatness-Based Adaptive Robust Control

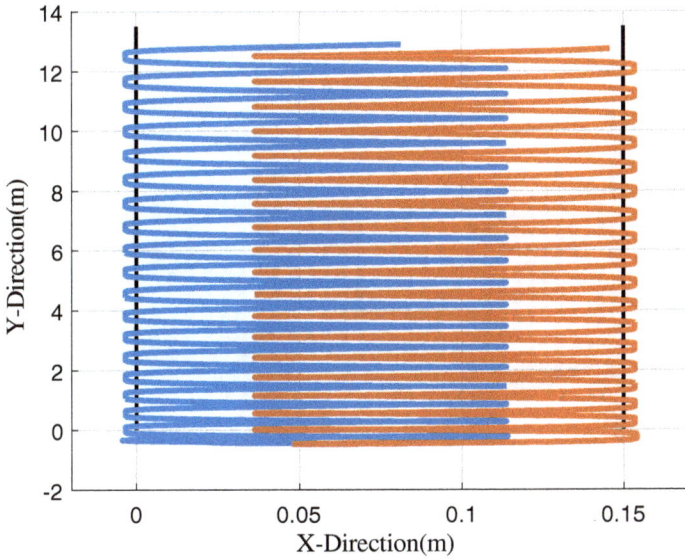

Fig. 6.10 Second link contact trajectory

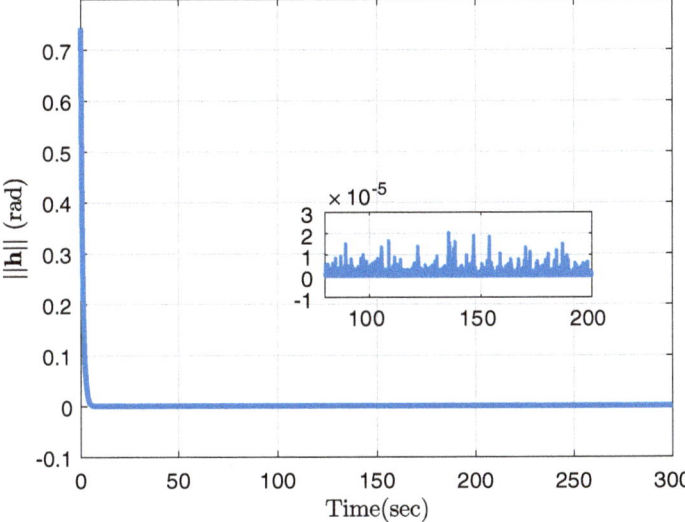

Fig. 6.11 Norm of VHCs

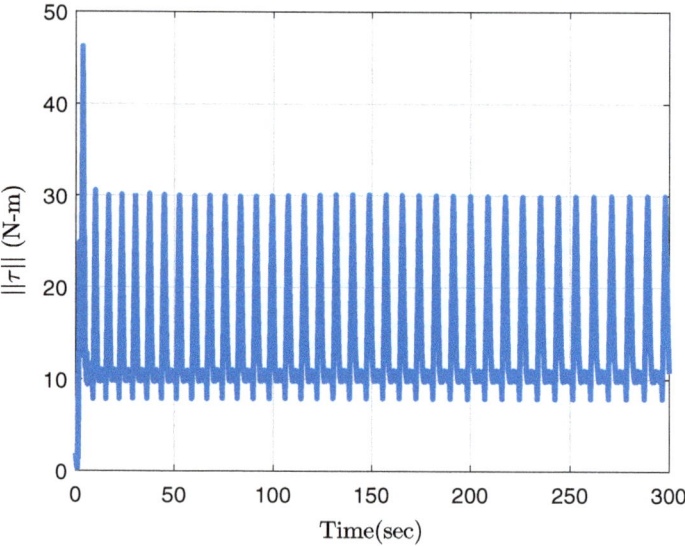

Fig. 6.12 Norm of control input

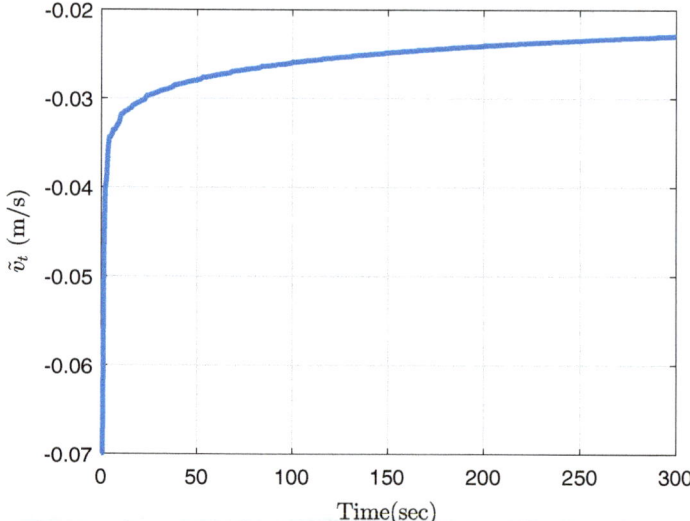

Fig. 6.13 Tangential velocity error

6.3 Flatness-Based Adaptive Robust Control

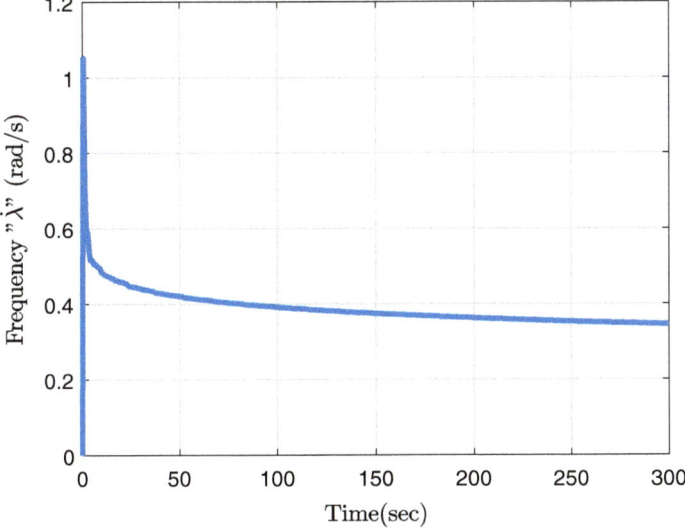

Fig. 6.14 Gait function frequency

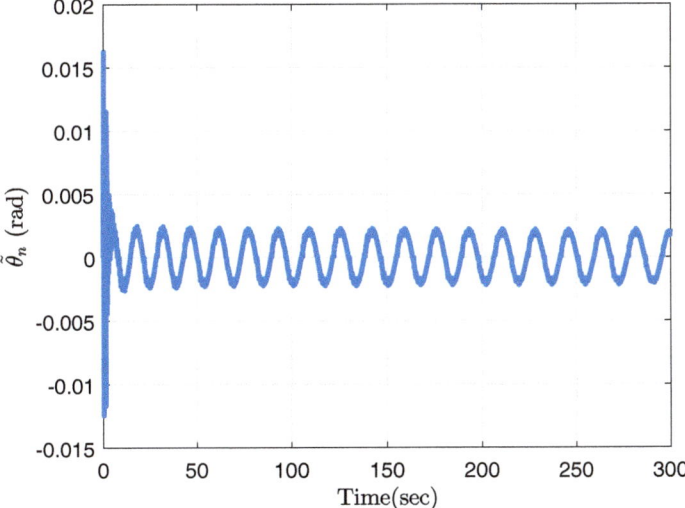

Fig. 6.15 Global head-angle error

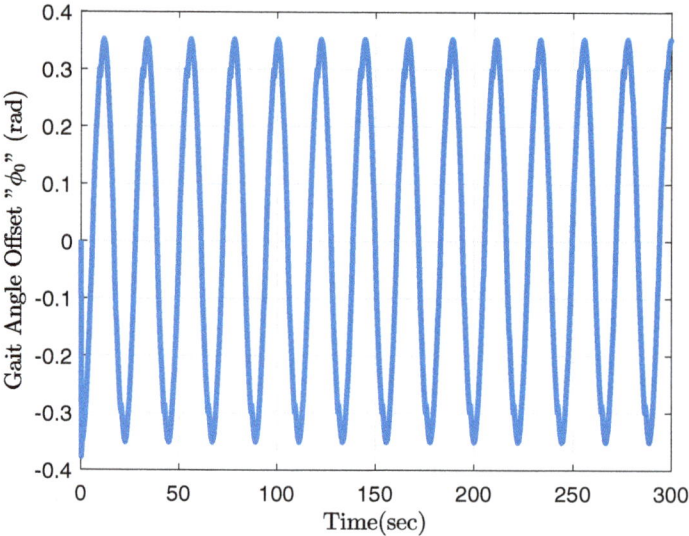

Fig. 6.16 Gait function offset

external disturbances which can deteriorate the tracking performance. It is essential to address such uncertainties through control law design to achieve the desired tracking behavior. To address these uncertainties during the in-channel motion, the differential-flatness-based adaptive robust controller presented in Sect. 5.5 has been employed.

6.3.1 Flat System

The approach can be applied to the system (6.23), (6.24) with similar choice of flat outputs as

$$\bar{\mathbf{F}}_t = \dot{\mathbf{F}}_t + \mathbf{K}_t \mathbf{F}_t, \tag{6.36a}$$

$$\bar{F}_n = \dot{F}_n + K_n F_n, \tag{6.36b}$$

$$F_{n+1} = K_v \tilde{v}_t + \dot{\lambda}, \tag{6.36c}$$

where $\mathbf{F}_t = \mathbf{h}(\lambda, \phi_0, \boldsymbol{\theta})$ and $F_n = \tilde{\theta}_n$. Similar to Sect. 5.5, $\mathbf{K}_t$ is a positive definite gain matrix whereas K_n and K_v are positive scalar gains to be appropriately chosen. Such choices would ensure the convergence of $\mathbf{F}_t$, $\dot{\mathbf{F}}_t$, F_n and $\dot{F}_n$ as well as boundedness of $\tilde{v}_t$ for the convergence of $\bar{\mathbf{F}}_t$, $\bar{F}_n$ and F_{n+1}. With the flat output vector as

6.3 Flatness-Based Adaptive Robust Control

$\mathbf{F} = [\bar{\mathbf{F}}_t^T \ \bar{F}_n \ F_{n+1}]^T \in \mathbb{R}^{n+1}$ and input $\mathbf{u} = [\boldsymbol{\tau}^T \ u_\lambda \ u_{\phi_0}]^T \in \mathbb{R}^{n+1}$, the flat system can be expressed as

$$\dot{\mathbf{F}} = \mathbf{f}_F + \bar{\mathbf{g}}_F \mathbf{u}_F, \tag{6.37}$$

where

$$\mathbf{f}_F = \begin{bmatrix} \mathbf{DM}^{-1}\left(\mathbf{W}\dot{\theta}^2 + b\mathbf{S}_\theta \mathbf{NC}_{f,c} - b\mathbf{C}_\theta \mathbf{NC}_{f,y} - \mathbf{C}_{f,\tau} - \mu_\tau \dot{\theta}\right) - \mathbf{\Phi}''\dot{\lambda}^2 + \mathbf{K}_t \dot{\mathbf{F}}_t \\ \psi_1 - \ddot{\theta}_{ref} + K_n \dot{F}_n \\ K_v(f_2 - \dot{v}_{ref}) \end{bmatrix},$$

$$\mathbf{g}_F = \begin{bmatrix} (\mathbf{DM}^{-1}\mathbf{D}^T) & -\mathbf{b}_1 & -\mathbf{\Phi}' \\ \mathbf{0}_{1\times(n-1)} & \psi_3 & \psi_2 \\ \mathbf{0}_{1\times(n-1)} & 0 & 1 \end{bmatrix},$$

with

$$\psi_1 = -\frac{\mathbf{e}^T \mathbf{MH} \mathbf{\Phi}''(\lambda)\dot{\lambda}^2}{\mathbf{e}^T \mathbf{Me}} + \frac{1}{\mathbf{e}^T \mathbf{Me}} \mathbf{e}^T \left(\mathbf{W}\dot{\theta}^2 + b\mathbf{S}_\theta \mathbf{NC}_{f,c} - b\mathbf{C}_\theta \mathbf{NC}_{f,y} - \mathbf{C}_{f,\tau} - \mu_\tau \dot{\theta}\right),$$

$$\psi_2 = -\frac{\mathbf{e}^T \mathbf{MH} \mathbf{\Phi}'(\lambda)}{\mathbf{e}^T \mathbf{Me}},$$

$$\psi_3 = -\frac{\mathbf{e}^T \mathbf{MH} \mathbf{b}_1}{\mathbf{e}^T \mathbf{Me}}.$$

The non-singularity of input matrix $\mathbf{g}_F$ has already been proven in Sect. 5.5. Taking a similar approach, the input matrix $\bar{\mathbf{g}}_F$ and dynamic function $\mathbf{f}_F$ of the flat system (6.37) are partitioned into nominal and uncertain parts as

$$\bar{\mathbf{g}}_F = \hat{\mathbf{g}}_F + \tilde{\mathbf{g}}_F,$$
$$\mathbf{f}_F = \hat{\mathbf{f}}_F + \tilde{\mathbf{f}}_F.$$

Hence, the flat system can be expressed as

$$\hat{\mathbf{g}}_F \dot{\mathbf{F}} = \hat{\mathbf{g}}_F \hat{\mathbf{f}}_F + \mathbf{u}_F + \mathbf{d}, \tag{6.38}$$

where $\mathbf{d} = \tilde{\mathbf{g}}_F \hat{\mathbf{f}}_F + \hat{\mathbf{g}}_F \tilde{\mathbf{f}}_F + \tilde{\mathbf{g}}_F \tilde{\mathbf{f}}_F - \tilde{\mathbf{g}}_F \dot{\mathbf{F}}$ represents the uncertainty in the flat system.

6.3.2 Adaptive Robust Control Law

This subsection details the adaptive robust control law employed to achieve robustness for the uncertain flat system (6.38).

Assumption 6.3 Considering the uncertainties in the system dynamics to be time varying, the uncertainty vector can be assumed to be bounded by a constant, i.e.

$$\|\mathbf{d}\| \leq \rho, \tag{6.39}$$

for $\rho > 0$ being constant coefficients. □

For uncertainties satisfying Assumption 6.3, a switching-law-based robust control law for the flat system (6.38) can be designed as

$$\mathbf{u}_F = -\hat{\mathbf{g}}_F \hat{\mathbf{f}}_F - \mathbf{K}_F \mathbf{F} - \eta_0 \operatorname{sat}(\mathbf{F}), \tag{6.40}$$

where $\mathbf{K}_F > 0 \in \mathbb{R}^{(n+1)\times(n+1)}$ is a suitable positive definite feedback gain matrix to be appropriately chosen and η_0 is a positive switching gain to be obtained through an adaptation law to achieve robustness. A saturation function has been employed to induce the switching behavior which can be written as

$$\operatorname{sat}(\mathbf{F}) = \begin{cases} \frac{\mathbf{F}}{\|\mathbf{F}\|} & \text{for } \|\mathbf{F}\| \geq \epsilon_F, \\ \frac{\mathbf{F}}{\epsilon_F} & \text{otherwise.} \end{cases} \tag{6.41}$$

6.3.3 Adaptation Law

Like in Sect. 5.5, this subsection presents an adaptation law designed to update switching gain to achieve efficient tracking performance by alleviating the overestimation problem. Also, it is not practical to know the value of ρ a priori so that a suitable value of η_0 can be chosen. For this reason, the dual-rate adaptation law [16] can be expressed as

$$\dot{\eta}_0 = \begin{cases} \bar{\eta}_{0,i} \|\mathbf{F}\| & for \; \{\eta_0 > 0, \mathbf{F}^T \dot{\mathbf{F}} > 0\} \; or \; \{\eta_0 \leq 0\} \\ -0.5 \bar{\eta}_{0,i} \|\mathbf{F}\| & for \; \{\eta_0 > 0, \mathbf{F}^T \dot{\mathbf{F}} \leq 0\} \end{cases}, \tag{6.42}$$

where

$$i = \begin{cases} 1 & for \; \|\mathbf{F}\| \leq \delta \\ 2 & for \; \|\mathbf{F}\| > \delta \end{cases}, \; \text{and} \; \eta_0(t_0) > 0.$$

As has been previously discussed, choice of the parameter δ is crucial as it represents the vicinity of the equilibrium point and can influence the performance of the adaptation law. The dual-gains $\bar{\eta}_{0,2} > \bar{\eta}_{0,1} > 0$ of the adaptation law dictate the rate of

6.3 Flatness-Based Adaptive Robust Control

change of the switching gain η_0. The adaptation law imitates a *course and fine tuning* policy to update the switching gain in an efficient manner. The rate of decreasing the gain has been kept as half of the increasing rate to maintain the stability of the closed-loop system. The stability analysis follows similar methodology as given in Sect. 5.5.3.

6.3.4 Simulation Results

This section presents the simulation scenario and results for the motion of a snake robot inside a channel with uncertainties. The physical specifications of the system considered for simulation are same as given in Table 6.1. Uncertainties have been considered by assuming different values of some parameters like joint friction coefficient, elasticity modulus, Poisson's ration and friction coefficient between the link surface and the channel wall. The nominal values of these parameters are given in Table 6.3. The simulation was run for 300 s and the initial condition of the generalized coordinates and compensators states are set to $\theta(0) = \dot{\theta}(0) = \mathbf{p}(0) = \dot{\mathbf{p}}(0) = \mathbf{0}$ and $\lambda = \dot{\lambda} = \phi_0 = \dot{\phi}_0 = 0$. The controller gains utilized for the control law discussed in the previous section are shown in Table 6.4.

Table 6.3 Nominal parameters

Nominal parameter	Numerical value
$\hat{\mu}_\tau$	0.45
$\hat{E}_{rub}$	0.04 MPa
$\hat{v}_{rub}$	0.45
$\hat{\mu}_{fric}$	0.3 Ns/m

Table 6.4 Controller and adaptation law parameters

Controller parameter	Numerical value
K_{a_1}	20
K_{a_2}	20
K_{n_1}	20
K_{n_2}	20
K_{n+1}	20
K_v	30
ϵ_F	0.01
δ	0.01
$\eta_0(0)$	0.1
$\eta_{0,1}$	0.05
$\eta_{0,2}$	0.1

6.3.5 Discussions

The trajectory and of the robot CM in global frame and its velocity profile in the global X- and Y-directions are shown in Figs. 6.17 and 6.18, respectively. These show the satisfactory tracking performance of the proposed approach along with the bounded velocity in the normal direction and velocity tracking along the channel. The trajectories of the first and the second link CMs shown in Figs. 6.19 and 6.20, respectively, exhibit their smooth transition from one wall to the other. The virtual penetrations during contacts by the first two links with the left and right channel wall are shown in Figs. 6.21 and 6.22, respectively. The norm of the actuated flat output $\mathbf{F}_a$ is shown in Fig. 6.23 with a steady-state convergence in the range of 10^{-5}. The norm of the total control effort required for the motion is shown in Fig. 6.24. The error in the tangential velocity as well as the heading angle exhibiting satisfactory tracking performance are shown in Figs. 6.25 and 6.27, respectively. The corresponding gait frequency as well as the gait offset for attaining the desired tangential velocity and head angle are given in Figs. 6.26 and 6.28, respectively. The norm of the flat output and the variation of switching gain over time are shown in Figs. 6.29 and 6.30. The sustained oscillations in the various plots is due to the switching control law. One can also observe that the switching gain remains bounded and provides robustness toward the uncertainties (Fig. 6.30).

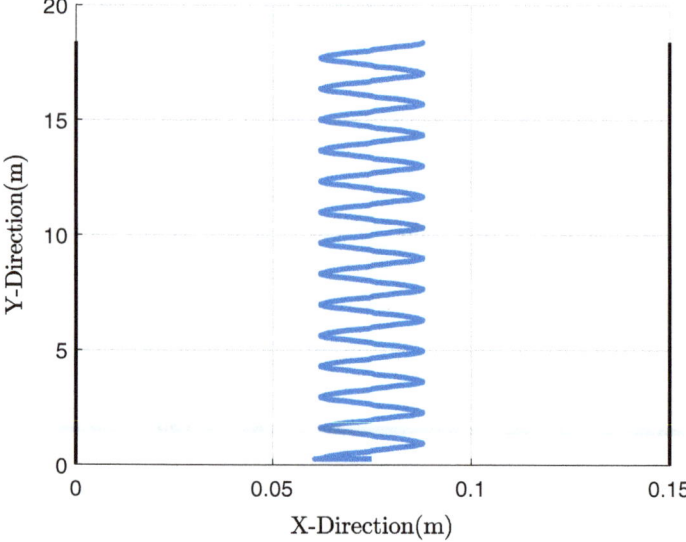

Fig. 6.17 Trajectory of robot CM

6.3 Flatness-Based Adaptive Robust Control

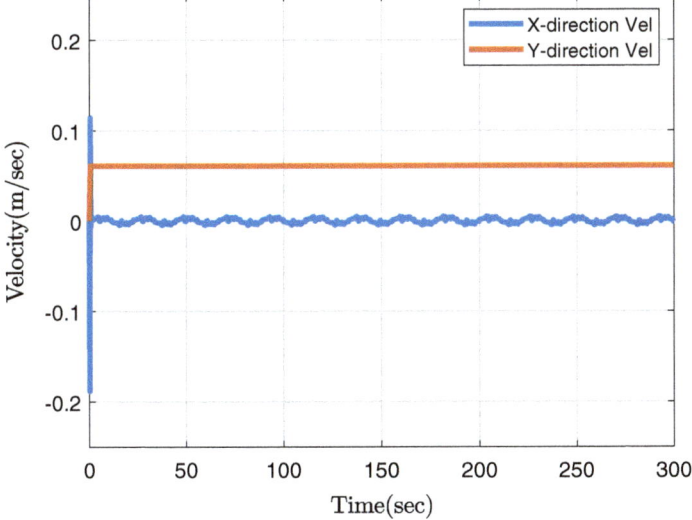

Fig. 6.18 Velocity profile of robot CM

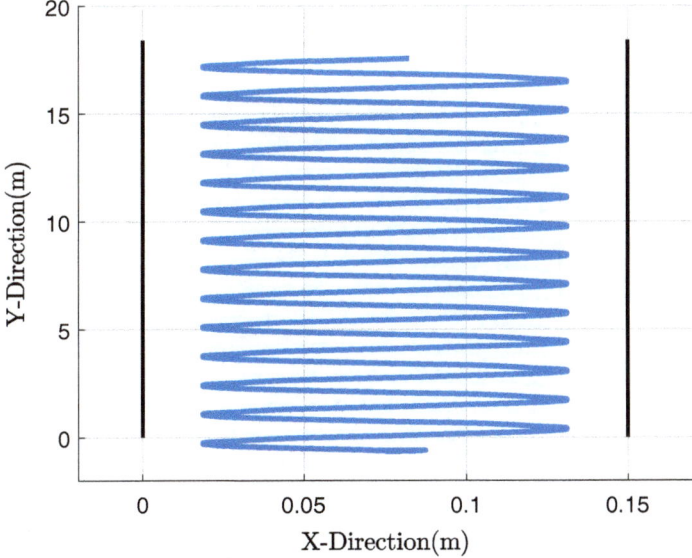

Fig. 6.19 First link CM trajectory

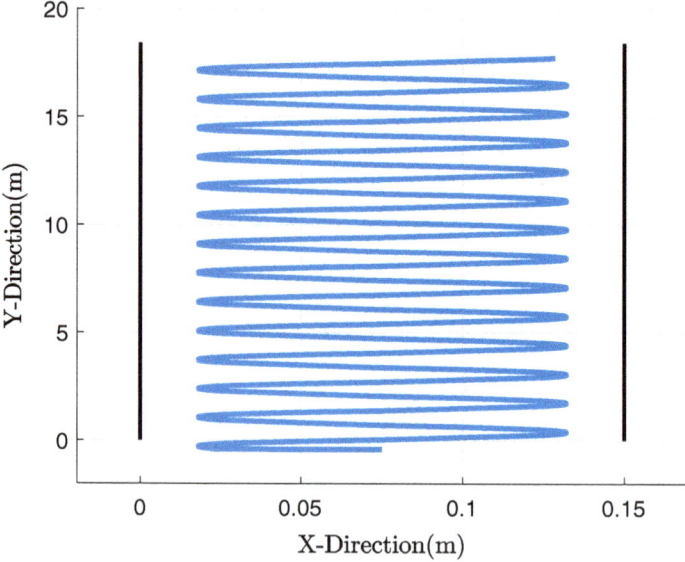

Fig. 6.20 Second link CM trajectory

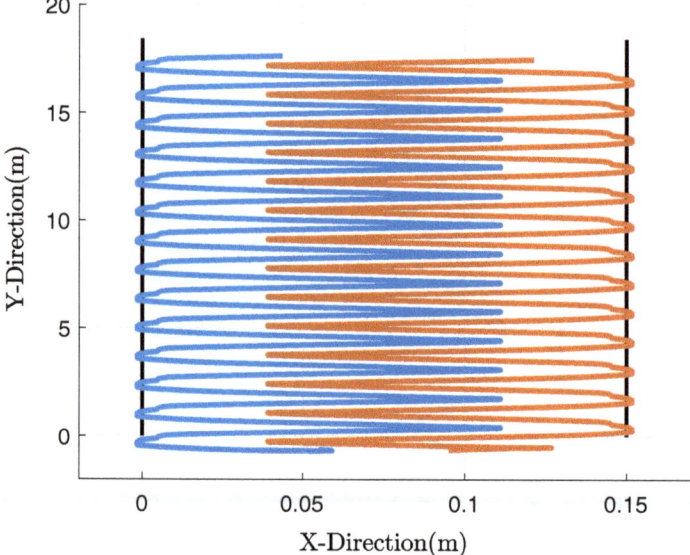

Fig. 6.21 First link contact trajectory

6.3 Flatness-Based Adaptive Robust Control

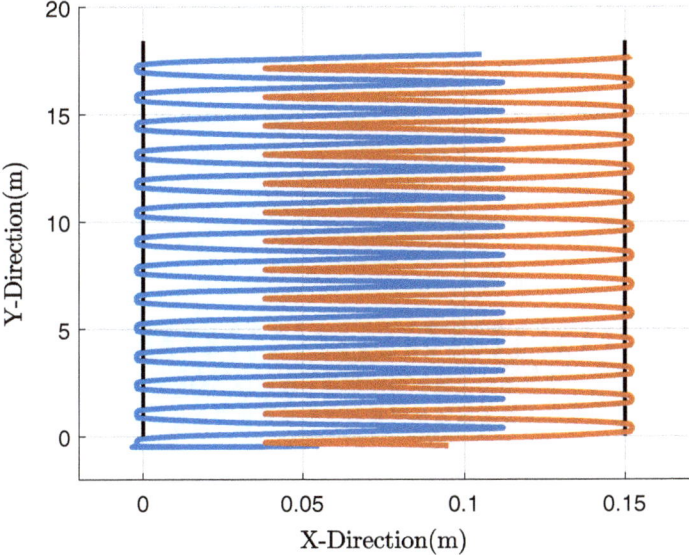

Fig. 6.22 Second link contact trajectory

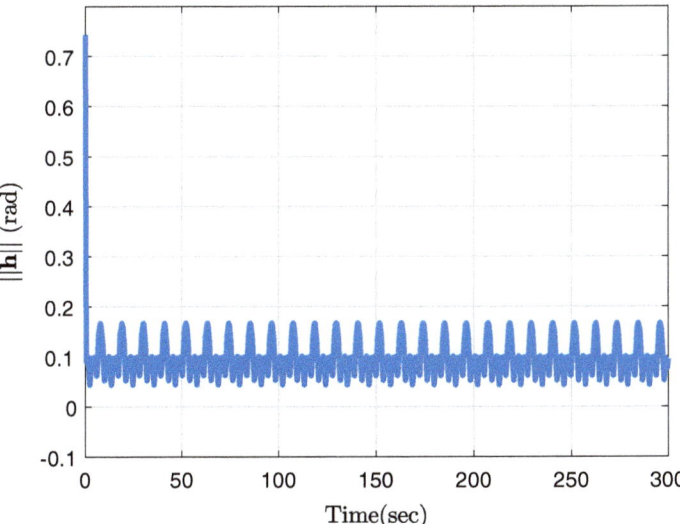

Fig. 6.23 Norm of VHCs

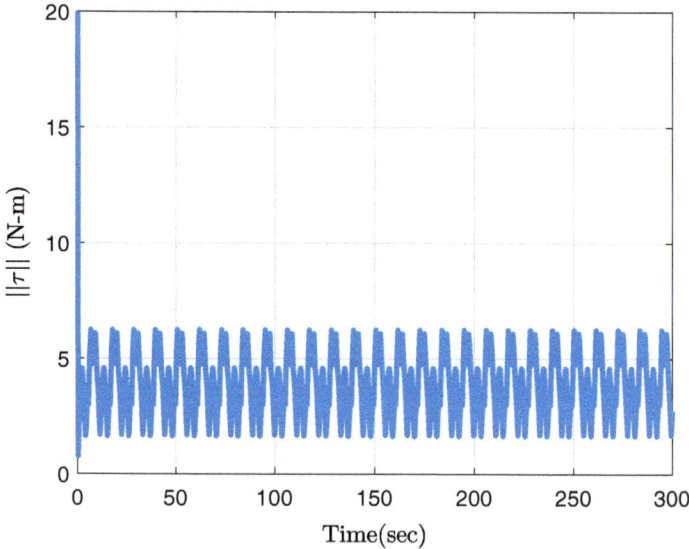

Fig. 6.24 Norm of control input

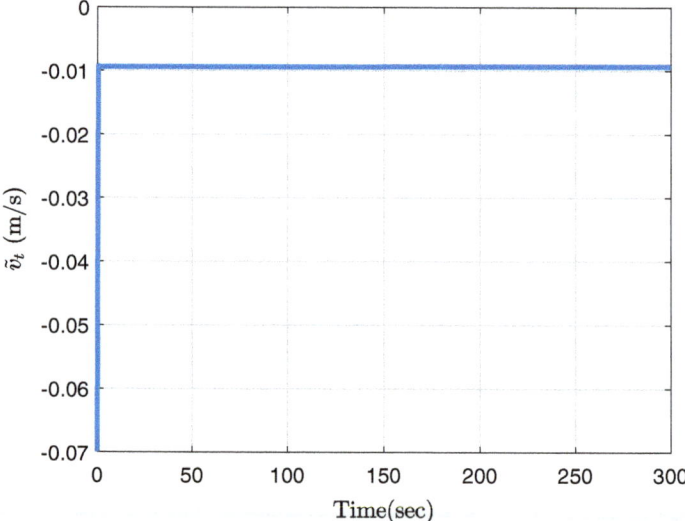

Fig. 6.25 Tangential velocity error

6.3 Flatness-Based Adaptive Robust Control

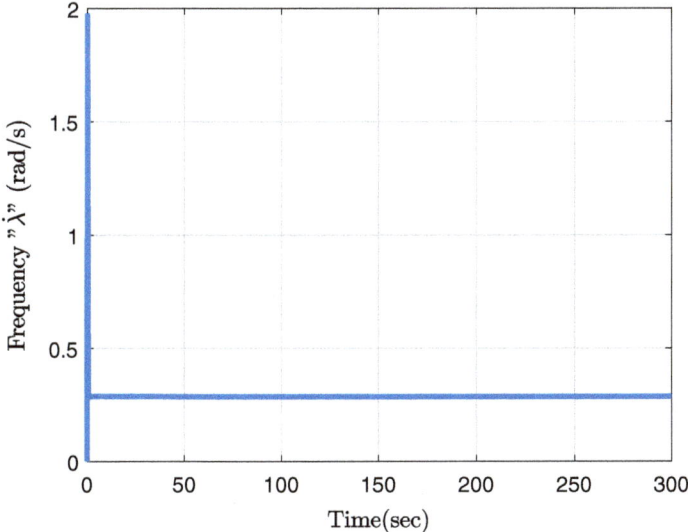

Fig. 6.26 Gait function frequency

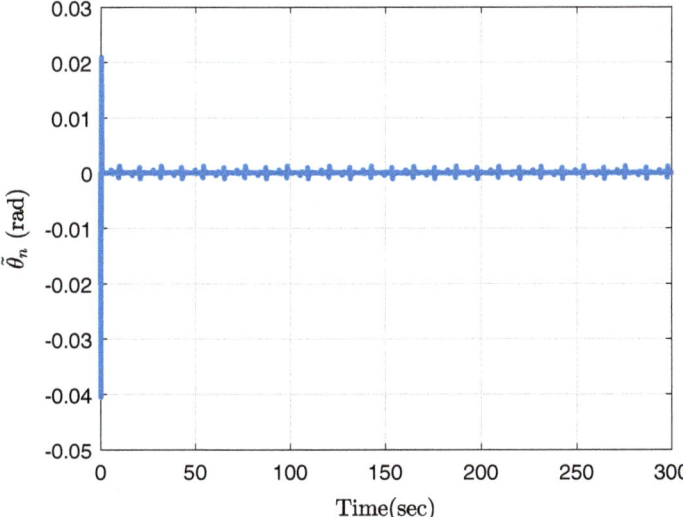

Fig. 6.27 Global head-angle error

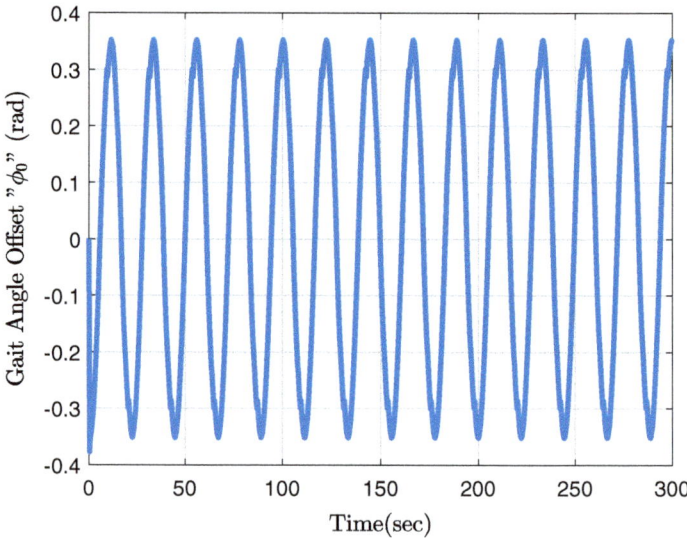

Fig. 6.28 Gait function offset

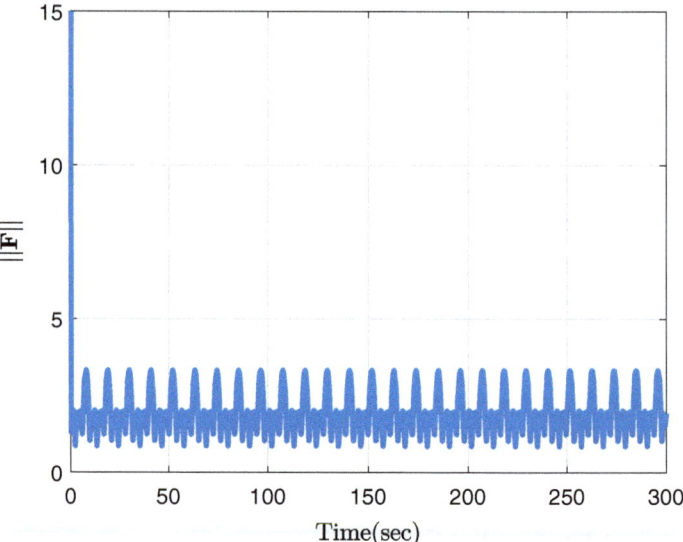

Fig. 6.29 Norm of flat output

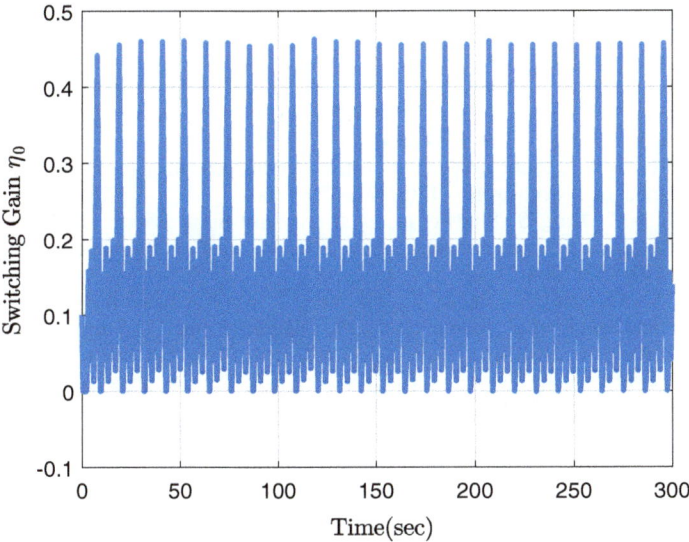

Fig. 6.30 Switching gain

6.4 Summary

This chapter discusses a method of modeling the dynamics of a snake robot inside a pipe. It has been assumed that the wall contact forces are in excess of those generated by friction from the floor. For each link, a contact force model has been proposed followed by a method to estimate the wall friction which generates the propagating force for the robot. The moment due to these force on the link CMs has been computed to write the moment balance equation. These one link models have been concatenated to generate the dynamic equations of the snake robot. A serpenoid gait motion has been imposed with amplitude depending upon some robot parameters and pipe dimensions. Control approaches for achieving desired tracking along the pipe has been described and the performance of the control laws have been verified through simulations. A flatness-based adaptive robust control approach has also been employed to address uncertainties during the in-channel motion of a snake robot. The actuator dynamics may be included in the snake dynamics so that overall controller design and analysis can be carried out. In this direction, the results of [17] in the context of mobile robot may be useful.

References

1. Brunete, A., Gambao, E., Torres, J., Hernando, M.: A 2 dof servomotor-based module for pipe inspection modular micro-robots. In: 2006 IEEE/RSJ International Conference on Intelligent Robots and Systems, pp. 1329–1334 (2006)
2. Choi, H., Ryew, S.: Robotic system with active steering capability for internal inspection of urban gas pipelines. Mechatronics **12**(5), 713–736 (2002)
3. Jun, C., Deng, Z., Jiang, S.: Study of locomotion control characteristics for six wheels driven in-pipe robot. In: IEEE International Conference on Robotics and Biomimetics, 2004. ROBIO 2004, pp. 119–124 (2004)
4. Fujiwara, S., Kanehara, R., Okada, T., Sanemori, T.: An articulated multi-vehicle robot for inspection and testing of pipeline interiors. Proc. IEEE/RSJ Int. Conf. Intell. Robots Syst. **1**, 509–516 (1993)
5. Okada, T., Sanemori, T.: Mogrer: a vehicle study and realization for in-pipe inspection tasks. IEEE J. Robot. Autom. **3**(6), 573–582 (1987)
6. Huang, H.P., Yan, J.L., Cheng, T.H.: Development and fuzzy control of a pipe inspection robot. IEEE Trans. Ind. Electron. **57**(3), 1088–1095 (2010)
7. Kim, J., Muramatsu, M., Murata, Y., Suga, Y.: Omnidirectional vision-based ego-pose estimation for an autonomous in-pipe mobile robot. Adv. Robot. **21**(3–4), 441–460 (2007)
8. Roh, S.G., Choi, H.R.: Differential-drive in-pipe robot for moving inside urban gas pipelines. IEEE Trans. Robot. **21**(1), 1–17 (2005)
9. Wakimoto, S., Suzumori, K., Takata, M., , Nakajima, J.: In-pipe inspection micro robot adaptable to changes in pipe diameter. J. Robot. Mechatron. **15**(6), 609–615 (2003). https://doi.org/10.20965/jrm.2003.p0609
10. Kuwada, A., Adomi, Y., Suzumori, K., Kanda, T., Wakimoto, S., Kadowaki, N.: Snake-like robot negotiating three-dimensional pipelines. In: Proceedings of the 2007 IEEE International Conference on Robotics and Biomimetics. Sanya, China (2007)
11. Johnson, K.L.: Contact Mechanics. Cambridge University Press, Cambridge (1985). https://doi.org/10.1017/CBO9781139171731
12. Liljeback, P., Pettersen, K.Y., Stavdahl, Ã., Gravdahl, J.T.: Controllability and stability analysis of planar snake robot locomotion. IEEE Trans. Autom. Control **56**(6), 1365–1380 (2011). https://doi.org/10.1109/TAC.2010.2088830
13. Liljebäck, P., Pettersen, K.Y., Stavdahl, Ø., Gravdahl, J.T.: Snake robots: modelling, mechatronics, and control. Springer Science & Business Media (2012)
14. Hirose, S., Morishima, A.: Design and control of a mobile robot with an articulated body. Int. J. Robot. Res. **9**(2), 99–114 (1990) https://doi.org/10.1177/027836499000900208
15. Mohammadi, A., Rezapour, E., Maggiore, M., Pettersen, K.Y.: Maneuvering control of planar snake robots using virtual holonomic constraints. IEEE Trans. Control Syst. Technol. **24**(3), 884–899 (2016). https://doi.org/10.1109/TCST.2015.2467208
16. Mukherjee, J., Roy, S., Kar, I.N., Mukherjee, S.: Maneuvering control of planar snake robot: an adaptive robust approach with artificial time delay. International Journal of Robust and Nonlinear Control (2021). https://doi.org/10.1002/rnc.5430 [In Press]
17. Das, T., Kar, I., Chaudhury, S.: Simple neuron-based adaptive controller for a nonholonomic mobile robot including actuator dynamics. Neurocomputing **69**(16), 2140–2151 (2006). https://doi.org/10.1016/j.neucom.2005.09.013. Brain Inspired Cognitive Systems

Chapter 7
Conclusions

Serpentine robots are designed to incorporate features of snake motion to achieve movement capabilities in wide range of environment like flat surfaces, constrained spaces, underwater, etc. Execution of motion control for such robots require the tracking of the tangential velocity to a predefined reference value, whereas the rest of the generalized coordinates are required to exhibit stable limit cycle behavior. This thesis contributes toward addressing maneuvering control problem as well as dynamic modeling for snake robots. Specifically, the design of a series of robust controllers for a planar snake robot dynamics, modeled with system uncertainties has been discussed. This thesis also studies differential flatness for snake robots, with the aim of facilitating an unified approach to achieve body shape, head angle and velocity control. The work in this thesis attempts to expand the work zone of a snake robot to constrained spaces like pipes and channels. A mathematical model for the motion in such an environment has been proposed with the aim of control design. The work presented in each chapter has been summarized below.

The book has been organized through six chapters and their respective contents are outlined as:

Chapter 1 introduces the state of art in the field of snake robotics, especially for dynamic modeling and the employed control methodologies. Some of the basic concepts required to understand the control methodologies adopted in the state-of-the-art approaches have been detailed in this chapter. The scope of contribution in this field have been identified that has lead to the motivations of the work enclosed in this book. Further, the organization of work elements of each of the subsequent chapter have been enumerated laying out the assorted theme of the book.

Chapter 2 addresses the problem of achieving robustness through control of a planar snake robot with bounded time-varying uncertainty in friction coefficients through a sliding-mode control (SMC) law for velocity and head-angle tracking. Moreover, the problem of overestimation of the switching gain has been alleviated through an adaptation law. This approach also circumvents the requirement of the knowledge of uncertainty bounds for implementation of the controller.

Chapter 3 presents a dual-layered artificial time-delay based approach (TDC) for maneuvering control of planar snake robot with system uncertainties. The boundedness assumption for these uncertainties can be obviated through this framework, under the less limiting assumption of the uncertainties to be slowly varying. Time delay based strategy (TDE) has been employed to estimate the uncertainties and the state functions of the dynamic model to compute the robust control law. Stability analysis using Lyapunov's method yield UUB stability of the closed-loop system and also provides a framework toward effective gain tuning to improve the tracking accuracy.

Chapter 4 details a dual-layer adaptive robust time-delayed controller (ARTDC) for head angle and velocity control of a planar snake robot with uncertainties. An adaptive robust control (ARC) law has been employed in addition to the TDC methodology to alleviate the effect of the estimation error of the TDE on the tracking performance. The ARC includes a switching control law and a novel dual-rate adaptation law to achieve improved tracking also simultaneously obviating the problem of overestimation of the switching gain.

Chapter 5 discusses differential flatness by considering a motivating example of wheeled mobile robot. This approach results in to a reduced-order system that simplifies controller design and trajectory generation. Subsequently, an augmented flat system has been obtained for the planar snake robot utilizing serpenoid gait function which has further been utilized for controller design. This method results in an one shot controller design for the whole snake robot system. The flatness-based approach has been employed to propose an adaptive robust control law to address uncertainties in the planar snake robot.

Chapter 6 proposes a dynamic model for the motion of a snake robot inside a constrained space. In this mode of locomotion, the robot utilizes contacts from the wall to generate propagating force for motion. Condition for detecting contact in simulation and a model for the contact and friction forces have been proposed assuming elliptical link geometry of the snake robot. The dynamic model has been used to design the state-of-the-art control law presented in Chap. 1 to demonstrate tracking. Parametric uncertainties have been considered in the system model and the flatness-based adaptive robust controller proposed in Chap. 5 have been utilized to achieve robust tracking.

Appendix A
Boundedness of the Time-Delayed Estimation Error for Outer Layer Time-Delayed Control

The time-delayed estimation error can be defined from the closed-loop system (3.7) as

$$\boldsymbol{\xi}(t) = \ddot{\mathbf{h}}(t) - \bar{\mathbf{u}}_h(t) \tag{A.1}$$

$$\Rightarrow \boldsymbol{\xi}(t) = \ddot{\mathbf{h}}(t) + \mathbf{K}_P \mathbf{h}(t) + \mathbf{K}_D \dot{\mathbf{h}}(t) = \mathbf{M}_h^{-1}(\bar{\mathbf{f}}_h - \hat{\mathbf{f}}_h). \tag{A.2}$$

Utilizing (3.2), (3.10) can be written as

$$\boldsymbol{\tau} = \left(\mathbf{M}_h(t-\gamma) - \bar{\mathbf{M}}_h\right)\ddot{\mathbf{h}}(t-\gamma) - \mathbf{f}_h(t-\gamma) - \mathbf{f}_\phi(t) + \bar{\mathbf{M}}_h \bar{\mathbf{u}}_h. \tag{A.3}$$

The estimation error (A.1) can be modified as

$$\mathbf{M}_h(t)\boldsymbol{\xi}(t) = \mathbf{M}_h(t)\ddot{\mathbf{h}}(t) - \mathbf{M}_h(t)\bar{\mathbf{u}}_h(t). \tag{A.4}$$

Using the system dynamics in (3.2), we obtain

$$\mathbf{M}_h(t)\boldsymbol{\xi}(t) = \mathbf{f}_h(t) + \boldsymbol{\tau}(t) + \mathbf{f}_\phi(t) - \mathbf{M}_h(t)\bar{\mathbf{u}}_h(t). \tag{A.5}$$

Further, substituting $\boldsymbol{\tau}$ from (A.3)

$$\mathbf{M}_h(t)\boldsymbol{\xi}(t) = \left(\bar{\mathbf{M}}_h - \mathbf{M}_h(t)\right)\bar{\mathbf{u}}_h(t) \\ + \left(\mathbf{M}_h(t-\gamma) - \bar{\mathbf{M}}_h\right)\ddot{\mathbf{h}}(t-\gamma) + \mathbf{f}_h(t) - \mathbf{f}_h(t-\gamma). \tag{A.6}$$

Now, adding and subtracting a term $\left(\bar{\mathbf{M}}_h - \mathbf{M}_h(t)\right)\bar{\mathbf{u}}_h(t-\gamma)$ in the RHS

$$\mathbf{M}_h(t)\boldsymbol{\xi}(t) = \left(\bar{\mathbf{M}}_h - \mathbf{M}_h(t)\right)\left(\bar{\mathbf{u}}_h(t) - \bar{\mathbf{u}}_h(t-\gamma)\right) + \mathbf{f}_h(t) - \mathbf{f}_h(t-\gamma) \\ + \left(\mathbf{M}_h(t-\gamma) - \bar{\mathbf{M}}_h\right)\ddot{\mathbf{h}}(t-\gamma) + \left(\bar{\mathbf{M}}_h - \mathbf{M}_h(t)\right)\bar{\mathbf{u}}_h(t-\gamma). \tag{A.7}$$

Utilizing the identity (A.1) as $\bar{\mathbf{u}}_h(t)(t-\gamma) = \ddot{\mathbf{h}}(t-\gamma) - \boldsymbol{\xi}(t)(t-\gamma)$, we obtain

$$\mathbf{M}_h(t)\boldsymbol{\xi}(t) = \big(\mathbf{M}_h(t) - \bar{\mathbf{M}}_h\big)\boldsymbol{\xi}(t-\gamma) - \big(\mathbf{M}_h(t) - \bar{\mathbf{M}}_h\big)\big(\bar{\mathbf{u}}_h(t) - \bar{\mathbf{u}}_h(t-\gamma)\big)$$
$$+ \mathbf{f}_h(t) - \mathbf{f}_h(t-\gamma) - \big(\mathbf{M}_h(t) - \mathbf{M}_h(t-\gamma)\big)\ddot{\mathbf{h}}(t-\gamma). \quad (A.8)$$

For the sake of simplicity, let us assume γ to be the sampling time of the system. Thus, time t can be expressed as k instant and time $(t-\gamma)$ as $(k-1)$ instant. The above equation can be expressed as

$$\mathbf{M}_h(k)\boldsymbol{\xi}(k) = \big(\mathbf{M}_h(k) - \bar{\mathbf{M}}_h\big)\boldsymbol{\xi}(k-1) - \big(\mathbf{M}_h(k) - \bar{\mathbf{M}}_h\big)\big(\bar{\mathbf{u}}_h(k) - \bar{\mathbf{u}}_h(k-1)\big)$$
$$+ \mathbf{f}_h(k) - \mathbf{f}_h(k-1) - \big(\mathbf{M}_h(k) - \mathbf{M}_h(k-1)\big)\ddot{\mathbf{h}}(k-1). \quad (A.9)$$

The dynamics of the time-delayed estimation can be expressed as

$$\boldsymbol{\xi}(k) = \big(\mathbf{I}_{(n-1)} - \mathbf{M}_h^{-1}(k)\bar{\mathbf{M}}_h\big)\boldsymbol{\xi}(k-1) + \boldsymbol{\eta}_1(k-1)$$
$$- \big(\mathbf{I}_{(n-1)} - \mathbf{M}_h^{-1}(k)\bar{\mathbf{M}}_h\big)\boldsymbol{\eta}_2(k-1), \quad (A.10)$$

where

$$\boldsymbol{\eta}_1(k-1) = \mathbf{M}_h^{-1}(k)\big(\mathbf{f}_h(k) - \mathbf{f}_h(k-1)\big) - \big(\mathbf{I}_{(n-1)} - \mathbf{M}_h^{-1}(k)\mathbf{M}_h(k-1)\big)\ddot{\mathbf{h}}(k-1),$$
$$\boldsymbol{\eta}_2(k-1) = \bar{\mathbf{u}}_h(k) - \bar{\mathbf{u}}_h(k-1).$$

In the dynamics of the time-delayed estimation (A.10), $\boldsymbol{\eta}_1(k-1)$ and $\boldsymbol{\eta}_2(k-1)$ are considered to be external disturbances. Assuming that the disturbances are bounded, a condition to ensure the convergence of the estimation error $\boldsymbol{\epsilon}(k)$ is given as [1]

$$\big\|\big(\mathbf{I}_{(n-1)} - \mathbf{M}_h^{-1}(k)\bar{\mathbf{M}}_h\big)\big\| < 1. \quad (A.11)$$

Appendix B
Boundedness of the Time-Delayed Estimation Error for Inner Layer Time-Delayed Control

The time-delayed estimation error can be defined as

$$\epsilon(t) = \dot{\sigma}(t) - \bar{\mathbf{u}}(t) \tag{B.1}$$

$$\Rightarrow \epsilon(t) = \dot{\sigma}(t) + \mathbf{K}\sigma(t) = \bar{\mathbf{g}}_1^{-1}(\bar{\mathbf{f}}_1 - \hat{\mathbf{f}}_1). \tag{B.2}$$

Utilizing (3.15), (3.20) can be written as

$$\mathbf{u}_\sigma(t) = \bar{\mathbf{g}}_1 \bar{\mathbf{u}}(t) - \big(\bar{\mathbf{g}}_1 - \mathbf{g}_1(t-\gamma)\big)\dot{\sigma}(t-\gamma) - \bar{\mathbf{f}}_1(t-\gamma). \tag{B.3}$$

The estimation error can be written as

$$\mathbf{g}_1(t)\epsilon(t) = \mathbf{g}_1(t)\dot{\sigma}(t) - \mathbf{g}_1(t)\bar{\mathbf{u}}(t). \tag{B.4}$$

Using the system dynamics in (3.15), we obtain

$$\mathbf{g}_1(t)\epsilon(t) = \mathbf{f}_1(t) + \mathbf{u}_\sigma(t) - \mathbf{g}_1(t)\bar{\mathbf{u}}(t). \tag{B.5}$$

Further, substituting $\mathbf{u}(t)$ from (3.20)

$$\mathbf{g}_1(t)\epsilon(t) = \big(\bar{\mathbf{g}}_1 - \mathbf{g}_1(t)\big)\bar{\mathbf{u}}(t) - \big(\bar{\mathbf{g}}_1 - \mathbf{g}_1(t-\gamma)\big)\dot{\sigma}(t-\gamma) \\ + \mathbf{f}_1(t) - \mathbf{f}_1(t-\gamma). \tag{B.6}$$

Now, adding and subtracting a term $\big(\bar{\mathbf{g}}_1 - \mathbf{g}_1(t)\big)\bar{\mathbf{u}}(t-\gamma)$ in the RHS

$$\mathbf{g}_1(t)\epsilon(t) = \big(\bar{\mathbf{g}}_1 - \mathbf{g}_1(t)\big)\big(\bar{\mathbf{u}}(t) - \bar{\mathbf{u}}(t-\gamma)\big) + \mathbf{f}_1(t) - \mathbf{f}_1(t-\gamma) \\ - \big(\bar{\mathbf{g}}_1 - \mathbf{g}_1(t-\gamma)\big)\dot{\sigma}(t-\gamma) + \big(\bar{\mathbf{g}}_1 - \mathbf{g}_1(t)\big)\bar{\mathbf{u}}(t-\gamma). \tag{B.7}$$

Utilizing the identity (B.1) as $\bar{\mathbf{u}}(t-\gamma) = \dot{\sigma}(t-\gamma) - \epsilon(t-\gamma)$, we obtain

$$\mathbf{g}_1(t)\epsilon(t) = \big(\mathbf{g}_1(t) - \bar{\mathbf{g}}_1\big)\epsilon(t-\gamma) + \big(\bar{\mathbf{g}}_1 - \mathbf{g}_1(t)\big)\big(\bar{\mathbf{u}}(t) - \bar{\mathbf{u}}(t-\gamma)\big)$$
$$+ \big(\mathbf{g}_1(t-\gamma) - \mathbf{g}_1(t)\big)\dot{\sigma}(t-\gamma) + \mathbf{f}_1(t) - \mathbf{f}_1(t-\gamma). \quad \text{(B.8)}$$

For the sake of simplicity, let us assume γ to be the sampling time of the system. Thus, time t can be expressed as k instant and time $(t-\gamma)$ as $(k-1)$ instant. The above equation can be expressed as

$$\mathbf{g}_1(k)\epsilon(k) = \big(\mathbf{g}_1(k) - \bar{\mathbf{g}}_1\big)\epsilon(k-1) + \big(\bar{\mathbf{g}}_1 - \mathbf{g}_1(k)\big)\big(\bar{\mathbf{u}}(k) - \bar{\mathbf{u}}(k-1)\big)$$
$$+ \big(\mathbf{g}_1(k-1) - \mathbf{g}_1(k)\big)\dot{\sigma}(k-1) + \mathbf{f}_1(k) - \mathbf{f}_1(k-1). \quad \text{(B.9)}$$

The dynamics of the time-delayed estimation can be expressed as

$$\epsilon(k) = \big(\mathbf{I}_2 - \mathbf{g}_1^{-1}(k)\bar{\mathbf{g}}_1\big)\epsilon(k-1) + \bar{\eta}_1(k-1) - \big(\mathbf{I}_2 - \mathbf{g}_1^{-1}(k)\bar{\mathbf{g}}_1\big)\bar{\eta}_2(k-1), \quad \text{(B.10)}$$

where

$$\bar{\eta}_1(k-1) = \big(\mathbf{g}_1^{-1}(k)\mathbf{g}_1(k-1) - \mathbf{I}_2\big)\dot{\sigma}(k-1) + \mathbf{g}_1^{-1}(k)\big(\mathbf{f}_1(k) - \mathbf{f}_1(k-1)\big),$$
$$\bar{\eta}_2(k-1) = \bar{\mathbf{u}}(k) - \bar{\mathbf{u}}(k-1).$$

In the dynamics of the time-delayed estimation (B.10), $\bar{\eta}_1(k-1)$ and $\bar{\eta}_2(k-1)$ are considered to be external disturbances. Assuming that the disturbances are bounded, a condition to ensure the convergence of the estimation error $\epsilon(k)$ is given as [1]

$$||(\mathbf{I}_2 - \mathbf{g}_1^{-1}(k)\bar{\mathbf{g}}_1)|| < 1. \quad \text{(B.11)}$$

Appendix C
Boundedness of Switching Gains for Adaptive Robust Time-Delayed Control

C.1 Boundedness of Outer Layer Switching Gain η_g

Considering the Lyapunov function for the outer loop as

$$V_h = \frac{1}{2}\bar{\mathbf{h}}^T \mathbf{\Gamma} \bar{\mathbf{h}}. \tag{C.1}$$

Taking time derivative,

$$\dot{V}_h = -\frac{1}{2}\bar{\mathbf{h}}^T \mathbf{\Theta} \bar{\mathbf{h}} + \mathbf{s}^T \boldsymbol{\xi} - \eta_h \mathbf{s}^T \operatorname{sat}(\mathbf{s}).$$

For $\mathbf{s}^T \dot{\mathbf{s}} \leq 0$, the adaptation law further reduce the switching gain to a lower positive value. Hence, in this proof, we are going to focus only on $\mathbf{s}^T \dot{\mathbf{s}} > 0$ which renders increment in the gain η_h. The condition $\mathbf{s}^T \dot{\mathbf{s}} > 0$ implies that $\exists \varpi \in \mathbb{R}^+$ such that $\|\mathbf{s}\| \geq \varpi$.

Case 1: $\|\mathbf{s}\| > \delta_h$

The time derivative of Lyapunov function V_h becomes

$$\dot{V}_h \leq -\frac{1}{2}\lambda_{\min}(\mathbf{\Theta}) \|\bar{\mathbf{h}}\|^2 - (\eta_h - \eta_h^*)\|\mathbf{s}\|.$$

Since η_h^* is a finite value by the virtue of condition (4.10) and η_h is increasing, there exists a finite time when $\eta_h \geq eta_h^*$ will occur [2]. Then one has

$$\dot{V}_h - \frac{1}{2}\lambda_{\min}(\mathbf{\Theta}) \|\bar{\mathbf{h}}\|^2 < 0.$$

The above condition implies $\mathbf{s}^T\dot{\mathbf{s}} < 0$ will occur and η_h will start decreasing following the adaptive law (4.13). Hence, whenever $\eta_h \geq \eta_h^*$ is satisfied during $\|\mathbf{s}\| > \delta_h$, the gain will start to decrease.

Case 2: $\|\mathbf{s}\| \leq \delta_h$
The derivative of V_h w.r.t. time yields

$$\dot{V}_h \leq -\frac{1}{2}\lambda_{\min}(\boldsymbol{\Theta})\|\bar{\mathbf{h}}\|^2 + \|\mathbf{s}\|\eta_h^* - \eta_h\frac{\|\mathbf{s}\|^2}{\delta_h}$$

$$\leq -\frac{1}{2}\lambda_{\min}(\boldsymbol{\Theta})\|\bar{\mathbf{h}}\|^2 + \delta_h\eta_h^* - \eta_h\frac{\varpi^2}{\delta_h}.$$

Similar to the argument made in the previous case, $\dot{V}_h < 0$ will occur in a finite time whenever $\eta_h \geq \eta_h^*(\delta_h^2/\varpi^2)$ and consequently η_h will start decreasing during $\|\mathbf{s}\| \leq \delta_h$.

Observing the results obtained in theses two cases it can be inferred that $\exists \bar{\eta}_h \in \mathbb{R}^+$ such that

$$\eta_h \leq \bar{\eta}_h \triangleq \max\{\eta_h^*\ \eta_h^*(\delta_h^2/\varpi^2)\}.$$

Apart from proving boundedness of the gain η_h, the above results convey that the switching gain do start decreasing in a finite time, which is important to avoid overestimation problem of switching gain [2].

C.2 Boundedness of Inner Layer Switching Gain η_σ

Considering a Lyapunov function $V_\sigma = \frac{1}{2}\boldsymbol{\sigma}^T\boldsymbol{\sigma}$ and proceeding similar like the previous subsection, it can be shown that $\exists \bar{\eta}_\sigma, \varpi_1 \in \mathbb{R}^+$ such that

$$\eta_\sigma \leq \bar{\eta}_\sigma \triangleq \max\{\eta_\sigma^*\ \eta_\sigma^*(\delta_\sigma^2/\varpi_1^2)\},$$

where $\|\sigma\| \geq \varpi_1$ when η_σ increases.

References

1. Hsia, T.C., Gao, L.S.: Robot manipulator control using decentralized linear time-invariant time-delayed joint controllers. In: Proceedings, IEEE International Conference on Robotics and Automation, vol. 3, pp. 2070–2075 (1990). https://doi.org/10.1109/ROBOT.1990.126310
2. Roy, S., Roy, S.B., Kar, I.N.: Adaptive-robust control of Euler–Lagrange systems with linearly parametrizable uncertainty bound. IEEE Trans. Control Syst. Technol. 1–9 (2017). https://doi.org/10.1109/TCST.2017.2739107

Index

A
Actuated flat output, 105
Adaptation law, 30
Adaptive robust control, 73
Adaptive sliding-mode control, 30
Angular moments, 6
Anisotropic friction, 2, 6
Articulation mechanism, 6
Augmented flat output, 106

B
Biological snakes, 2
Biomimetics, 1
Body-shape control, 13, 137
Boundary layer dynamics, 18
Bounded uncertainty, 28

C
Center of mass, 4, 132
Centripetal and Coriolis matrix, 6
Closed loop system, 30, 58, 63, 76, 79, 99, 108
Compensator, 17
Constraint forces, 135
Contact force model, 129
Control problem of snake robots, 7
Coulomb friction model, 134

D
Differential flatness, 94
Differential weight, 94

Double pendulum, 11
Dual-rate adaptation law, 77, 80, 117
Dynamic equations, 6

E
Elliptical link profile, 129
Equivalent control, 29
Euler–Lagrangian system, 9
Exponentially stable, 10

F
Feedback control law, 108
Feedback controller, 16
Finite-time convergence, 37
Finite time reachability, 32
Finite-time stability, 33
Flatness-based ARC, 113, 142
Flat output, 97
Flat system, 97, 107
Free-body diagram, 130
Friction coefficient, 133
Friction force, 134
Friction force moment, 134

G
Gait function amplitude, 15
Gait function delay, 15
Gait function frequency, 15
Gait function offset, 15

H
Head-angle control, 18
Hertz theory, 131
Hurwitz matrix, 77

I
Inertia matrix, 6
Inner layer ARTDC, 78
Inner layer TDC, 59
Input-output linearization, 19

J
Joint viscous damping, 135

K
Kinematics, 5

L
Lateral undulation, 6
Link polar coordinate, 129
Logistic function, 131
Lyapunov equation, 76
Lyapunov stability analysis, 32, 38, 80, 99, 109, 118

M
Modulus of elasticity, 131

N
Newton–Euler dynamics, 135
Nonholonomic constraint, 96
Nonsingularity of input matrix, 33
Normal force moment, 134
Normal velocity, 19
Numerical simulations, 40

O
Outer layer ARTDC, 74
Outer layer TDC, 57
Overactuated system, 107

P
Penetration, 129
Physical specifications, 40
Poisson's ratio, 131

Positive definiteness, 77
Posture-error, 97
Progressive phase delay, 6
Prolongation, 95
Pseudo-inputs, 106

R
Radius of curvature, 131
Real sliding mode, 30, 38
Reduced order dynamics, 29
Robust control law, 116

S
Saturation function, 76
Serpenoid curve, 14
Serpenoid gait function, 14, 136
Singularly perturbed system, 18
Sliding-mode control, 29
Sliding surface, 29
Slowly varying uncertainties, 73
Smooth geometry, 131
Snake robot, 1
Stability analysis, 63
Swimming snake robot, 2
Switching control, 29
Switching gain, 32
Switching gain overestimation, 30

T
Tangential velocity, 19
Time-delayed control, 58
Time-delayed estimation, 58, 75, 76
Time varying friction, 40
Torque balance, 135

U
Underactuation, 9
UUB stability, 64, 81, 83

V
Velocity control, 19
VHC generator, 11
Virtual holonomic constraints, 8
Viscous friction model, 5

W
Wheeled mobile robot, 95

Lightning Source UK Ltd.
Milton Keynes UK
UKHW020737170522
403097UK00002B/6